Understanding Biomass Energy

Understanding Biomass Energy

Mark Cook

MURPHY & MOORE
www.murphy-moorepublishing.com

Published by Murphy & Moore Publishing,
1 Rockefeller Plaza,
New York City, NY 10020, USA
www.murphy-moorepublishing.com

Understanding Biomass Energy
Mark Cook

International Standard Book Number: 978-1-63987-546-7 (Hardback)

Cataloging-in-Publication Data

Understanding biomass energy / Mark Cook.
p. cm.
Includes bibliographical references and index.
ISBN 978-1-63987-546-7
1. Biomass energy. 2. Energy conversion. 3. Fuel. 4. Refuse as fuel. I. Cook, Mark.
TP339 .U53 2022
662.88--dc23

TABLE OF CONTENTS

PREFACE

Any plant or animal material that can be used to produce energy in the form of electricity or heat is called biomass. These are organic materials which store chemical energy sourced from the Sun. Wood and its residues are the most widely used biomass materials. They can be used either directly or can be processed into pellet form. Some of the other examples of biomass feedstock are corn, bamboo, switchgrass and miscanthus. Waste materials such as municipal waste, wood waste, agricultural waste, sewer sludge and manufacturing wastes also find application as sources of energy. Biomass conversion processes upgrade raw biomass into higher grade fuels. These methods can be classified into thermal conversions, chemical conversions, biochemical conversions and electrochemical conversions. Some of the issues related to the use of biomass are emission of greenhouse gases, forest conservation and carbon payback time. Biomass energy is an upcoming field of science that has undergone rapid development over the past few decades. This book unfolds the innovative aspects of biomass energy which will be crucial for the holistic understanding of the subject matter. It will serve as a valuable source of reference for those interested in this field.

To facilitate a deeper understanding of the contents of this book a short introduction of every chapter is written below:

Chapter 1- The fuel which is developed from organic materials and is used to generate electricity and other forms of power is termed as biomass. It is a renewable and sustainable source of energy. It is broadly categorized into two types, namely, waste biomass and energy crops. This chapter provides a brief introduction to these types of biomass as well as biofuels and biomass composites.

Chapter 2- Biomass conversion involves the usage of heat or thermal energy for the conversion of biomass energy into another form of chemical energy. Biomass heating systems replace fossil fuels for in different types of heating systems such as radiators and warm air heating systems. The topics elaborated in this chapter will help in gaining a better perspective about biomass conversion and biomass heating systems.

Chapter 3- The fuel which is derived through various processes from biomass is termed as biofuel. Some of the prominent types of biofuels are bioethanol, biodiesel, green diesel and aviation biofuel. This chapter closely examines the key aspects of these types of biofuels as well as hydrogen production using microbial biomass.

Chapter 4- Biofuels can be sourced from a variety of sources such as plants and different types of wastes. A few of these sources are corn, sugarcane, vegetable oil, cellulose, algal oil, camelina, jatropha, animal fat, paper waste and rapeseed. The topics elaborated in this chapter will help in gaining a better perspective about these sources of biofuels.

Chapter 5- Biomass finds applications for numerous purposes such as for the production of biogas, biochar, bioliquids, blue carbon and bioasphalt. It is also used for cogeneration and production of mixed alcohol fuels through bioconversion. These diverse applications of biomass have been thoroughly discussed in this chapter.

Finally, I would like to thank the entire team involved in the inception of this book for their valuable time and contribution. This book would not have been possible without their efforts. I would also like to thank my friends and family for their constant support.

Mark Cook

Chapter 1

Understanding Biomass

The fuel which is developed from organic materials and is used to generate electricity and other forms of power is termed as biomass. It is a renewable and sustainable source of energy. It is broadly categorized into two types, namely, waste biomass and energy crops. This chapter provides a brief introduction to these types of biomass as well as biofuels and biomass composites.

Biomass is the material derived from plants that use sunlight to grow which include plant and animal material such as wood from forests, material left over from agricultural and forestry processes, and organic industrial, human and animal wastes. Biomass energy is a type of renewable energy generated from biological (such as, anaerobic digestion) or thermal conversion (for example, combustion) of biomass resources. Biomass comes from a variety of sources which include:

- Wood from natural forests and woodlands.
- Forestry plantations.
- Forestry residues.
- Agricultural residues such as straw, stover, cane trash and green agricultural wastes.
- Agro-industrial wastes, such as sugarcane bagasse and rice husk.
- Animal wastes.
- Industrial wastes, such as black liquor from paper manufacturing.
- Sewage.
- Municipal solid wastes (MSW).
- Food processing wastes.

In nature, if biomass is left lying around on the ground it will break down over a long period of time, releasing carbon dioxide and its store of energy slowly. By burning biomass its store of energy is released quickly and often in a useful way. So converting biomass into useful energy imitates the natural processes but at a faster rate.

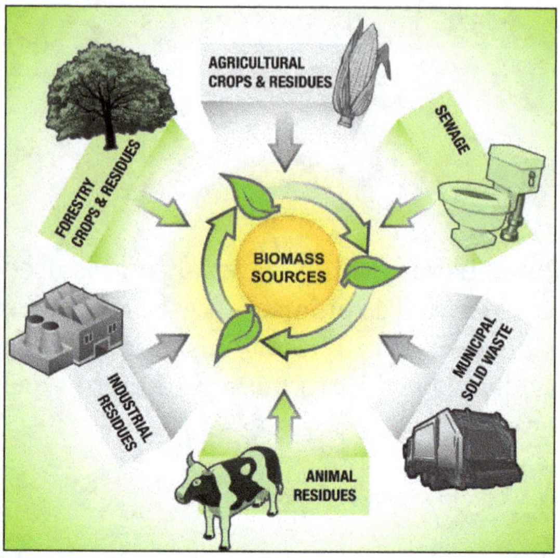

Biomass can be transformed into clean energy and/or fuels by a variety of technologies, ranging from conventional combustion process to advanced biofuels technology. Besides recovery of substantial energy, these technologies can lead to a substantial reduction in the overall biomass waste quantities requiring final disposal, which can be better managed for safe disposal in a controlled manner while meeting the pollution control standards.

Biomass conversion system reduces greenhouse gas emissions in two ways. Heat and electrical energy is generated which reduces the dependence on power plants based on fossil fuels. The greenhouse gas emissions are significantly reduced by preventing methane emissions from decaying biomass. Moreover, biomass energy plants are highly efficient in harnessing the untapped sources of energy from biomass resources and helpful in development of rural areas.

Availability of Biomass

Biomass is one of the most available renewable energy sources on our planet. In 2011 biomass satisfies around 10% of the primary energy use of the world. Nevertheless, its use is not shared in a homogeneous way. In developing countries, biomass covers from 34 to 40% of the total energy need. On the other hand, in industrialised countries its contribution is very limited and biomass only amount to 3% of primary energy use. In particular, the United States obtain 3.2% of their energy from biomass, Europe obtains a total of 3.5% of its energy from biomass. In Sweden and Finland, the electric energy obtained by burning the waste of the forest industry satisfies respectively 17% and 15% of the domestic demand for electric energy. In Austria, biomass covers 13% of the domestic electric need.

Instead, the contribution of biomass to the Italian national balance is limited to 2-3%. That distribution does not only take into account the attention and economic effort shown by certain countries in terms of search for and development of new technologies, but also the actual availability of land where "energy cultures" are to be implemented. Estimates show that, in developed countries, replacing fossil fuels with biomass would require over 950 mega-hectares (millions of hectares) of land available to be devoted to energy cultures. The countries in the northern central Europe are extremely developed as far as the use of biomass as energy source is concerned. They have installed large co-generation (joint production of electric energy and heat) and biomass-supplied district heating plants. France, that has the widest agricultural surface in Europe, invests a lot in the production of bio-diesel and ethanol, for which a de-taxation policy has been adopted.

The UK, instead, has developed limited production of bio-fuels, which are currently considered as too expensive, and has dedicated to the development of a large and efficient system for the recovery of biogas from dumps, both for thermal and electric aims. Sweden and Austria, that boast a long tradition of firewood, have continued to increase the use of firewood for heating and district heating, encouraging the growth of copses (willow tree, poplars), that provide a yield that is 3-4 times higher than the average supply of these raw materials.

Fire, unquestionably the most important discovery in the history of mankind was discovered thanks to the accidental combustion of wood. Fire has illuminated, heated, protected and fed mankind for thousands of years. Briefly, fired fostered the birth of civilisation. Wood, on the other hand, remained the most widely used raw material for many centuries, not only to burn fires, but also as building material. The invention of the steam engine allowed mankind to obtain mechanic energy from the combustion of wood, whereas up to the 18th century wind and water where the only mechanic energy sources available, thanks to wind and water mills.

During the Industrial Revolution wood started to become scarce owing to the massive deforestation carried out to produce energy. Mankind had to look for alternative energy sources and found them in coal and oil, which at the time were abundant albeit not renewable. Only recently, energy need and the possible disappearance of fossil fuels and the pollution produced by the combustion led man to "re-discover" the usefulness of wood and biomass as energy sources.

Boost to Biomass

In the past 10 years, there has been renewed interest, world-wide, in biomass as an energy source. There are several reasons for this situation:

- Firstly, technological developments relating to the conversion, crop production, etc. promise the application of biomass at lower cost and with higher conversion efficiency than was possible previously. For example, when low cost biomass residues are used for fuel, the cost of electricity is already now often competitive with fossil fuel-based power generation. More advanced options to produce electricity are looking promising and allow a cost-effective use of energy crops e.g. production of methanol and hydrogen by means of gasification processes.

- The second main stimulus is the agricultural sector in Western Europe and in the US, which is producing food surpluses. This situation has led to a policy in which land is set aside in order to reduce surpluses. Related problems, such as the de-population of rural areas and payment of significant subsidies to keep land fallow, makes the introduction of alternative, non-food crops desirable. Demand for energy will provide an almost infinite market for energy crops grown on such (potentially) surplus land.
- Thirdly, the potential threat posed by climate change, due to high emission levels of greenhouse gases (CO_2 being the most important one), has become a major stimulus for renewable energy sources in general. When produced by sustainable means, biomass emits roughly the same amount of carbon during conversion as is taken up during plant growth. The use of biomass therefore does not contribute to a buildup of CO_2 in the atmosphere.

But these three main issues are not the only stimuli: biomass is also an indigenous energy source, available in most countries and its application may diversify the fuel-supply in many situations, which in turn may lead to a more secure energy supply. Biomass production can generate employment and if intensive agriculture is replaced by less intensively managed energy crops, there are likely to be environmental benefits, such as reduced leaching of fertilisers and reduced use of pesticides. Moreover, if appropriate crops are selected, restoration of degraded lands may be possible. Depending on the crops used and the way the biomass is cultivated, increased biodiversity can be obtained, compared to current agricultural practice.

Biomass is available on a renewable basis, either through natural processes, or it can be made available as a by-product of human activities i.e. organic wastes. The potential of biomass energy derived from forest and agricultural residues worldwide, is estimated at about 30 EJ/yr compared to an annual world-wide energy demand of over 400 EJ. If biomass is to contribute to a larger extent to the world's energy supply, then energy farming, the cultivation of dedicated crops for energy purposes, will be required, using fallow land and marginal lands, the latter being largely unsuited for food crops.

When energy crops are considered as a source of biomass, the total energy potential of biomass for energy production may be considerably larger than the energy potential of biomass residues. In 1992 at the Rio United Nations Conference on environment and development, the renewable intensive global energy scenario (RIGES) suggested that, by 2050, approximately half the world's current primary energy consumption of about

400 EJ/yr, could be met by biomass and that 60% of the world's electricity market could be supplied by renewables, of which biomass is a significant component. In the UK, the Government's target is to generate 10% of the national electricity need of 60 GW/yr from renewable sources, of which biomass will form a significant part. To date, the 10 or so biomass projects working/under construction will generate about 100 MW. Biomass can be converted into three main types of product:

- Electrical/heat energy.
- Transport fuel.
- Chemical feedstock.

Types of Biomass

Biomass can be split into two distinct categories waste biomass and energy crops.

Waste Biomass

Forestry Residue

Forestry residues include biomass not harvested or removed from logging sites in commercial forests, as well as material resulting from forest management operations such as pre-commercial thinnings and removal of dead and dying trees. Wood is the most commonly used biomass fuel for heat and power, and unwanted woody waste is a by-product of many forestry operations. Using these materials for electricity generation recovers their energy value while avoiding landfill disposal.

Forest Residue.

Forestry waste includes logging residues, imperfect commercial trees, dead wood, and other non-commercial trees that need to be thinned from crowded, unhealthy, fire-prone forests. Forest thinning is necessary to help some forests regain their natural health, but for smaller woodlands the cost of removing the wood cannot be recovered through timber sales due to their poor quality. Commercial amounts of wood for fuel can be produced as a by-product of forestry management. Wood has a relatively low calorific value of around 19GJ/dry tonne. When harvested, wood has moisture content of around 55% by weight. In Scotland there are over 1,318,000 hectares of woodland and forest, and is estimated to generate sustainable wood fuel for nearly 250,000 homes.

Substantial areas of forest are needed to support a wood-fired power station. For example, a 6MW station with an efficiency of 20% would need between 430 and 2,150 hectares a year of sustainably managed forest at harvest. This exceptionally wide range shows the variable yield of forest residues. The yield itself depends on various factors such as terrain, accessibility, tree species and age, and end use of the timber as opposed to the residues will be put. However, the feasibility of this will also depend on other issues, such as the conservation status of the woodland, accessibility and countryside policies.

Animal Farming

Poultry Litter

Poultry litter is the bedding material collected from broiler sheds. It's usually made up of wood shavings, shredded paper or straw, mixed with the chicken droppings. It has a calorific value of 9-15GJ/tonne, which is slightly lower than that for wood. It has a moisture content of between 20-50% depending on the methods of husbandry used by farmers.

An example of using poultry litter as a biomass fuel source can be found in Fife. The Westfield biomass plant uses a combustion system to burn poultry litter to generate electricity and produce fertiliser. It has a net electricity output of 10MW and has been fully operational since October 2000. The practice of spreading large quantities of poultry litter on the land is no longer considered acceptable because it can cause serious environmental problems by polluting watercourses and producing odours if not correctly managed. The poultry farming industry has come under increasing pressure from environmental agencies to adopt a more environmentally acceptable method of disposal.

Cattle Farming

Cattle farming techniques significantly affect the quantity and quality of manure that may be delivered to the anaerobic digestion system. The number of cows, the housing, transport, and bedding systems used by the farms determines the amount of slurry that must be used and therefore the amount of energy produced.

Free Stall Barn.

Cattle farming may be housed using a variety of methods. The most commonly used systems include free stalls, corrals with paved feed lanes, and open lot systems. The type of housing used determines the quantity and quality of manure that can be economically collected. The commonly used methods of collecting the manure are flush, scrape, vacuum, and loader systems. In free stall barns the manure can be flushed, scraped, or vacuum collected.

Pig Farming

Intensive Pig Farming.

Several options for collecting and storing swine manure are available, depending on the manure form. Common storage methods include underfloor pits, outdoor (above or below ground) structures, earthen pits, lagoons and holding ponds. Flushing gutters and scraper systems are among the methods used to collect and transport manure to appropriate storage facilities. As with cattle farming, the manure can be used as a fuel source for anaerobic digestion.

Slaughterhouse and Fishery Waste

At a slaughterhouse or a fish processing plant, there is a huge amount of organic waste. This has the possibility of being a danger to the environment and human or animal health. The EU Animal By-Products Regulation specifies these animal wastes may be disposed of safely. This can be an extremely costly process, but this type of waste can also be used as a feed stream for Anaerobic Digestion.

Organic Municipal Solid Waste (MSW)

Organic MSW is any matter collected from commercial or residential properties such as food waste, paper etc. Organic waste, whether from commercial or residential properties, makes up a substantial amount of waste that is landfilled. As with other wastes, it can converted into energy by various methods. One is direct combustion (incinerator), or by anaerobic digestion in a landfill or in a process plant.

Sewage Waste

Sewage waste is a source of biomass that is comparable to the other animal wastes. Energy can be extracted from sewage using anaerobic digestion, pyrolysis or drying and incineration.

Energy Crops

Energy crops are produced specifically for their fuel value. There are several crops that can be readily utilised as energy sources. Both Short Rotation Coppice (SRC) and Miscanthus are covered under the Energy Crops Scheme (ECS).

Short Rotation Coppice (SRC)

SRC is a densely planted, high-yielding varieties of either willow or poplar, harvested on a 2 to 5 year cycle, although commonly every 3 years. SRC is a woody, perennial crop, the rootstock or stools remaining in the ground after harvest with new shoots emerging the following spring. A plantation could last up to 30 years before re-planting becomes necessary, although this depends on the productivity of the stools.

Short rotation crop field.

In the UK, yields achievable from willow SRC at first harvest are expected to be in the range 7 to 12 oven dry tonnes per hectare per year depending on site and efficiency of establishment. New varieties are expected to greatly increase yields. It would take roughly 3,000 ha of SRC to sustain a 5 MWe gasifier.

Miscanthus

Miscanthus crops are woody, perennial, rhizomatous grasses, which can be harvested annually for at least 15 years. By the third year harvestable yields are between 10-13 tonnes per hectare. Peak harvestable yields of 20 tonnes per hectare have been recorded. Defra Energy Crops 22,000 tonnes can provide enough electricity to power 2,000 homes.

Pellets

Pellets are a refined, solid fuel biomass with a low moisture content, which makes it easy to transport, store and convert into energy. It is manufactured from saw dust, wood chips, shaving or bark. Pellets are typically 6-8mm in diameter and 5-30 mm long. The maximum water content is 8%. The following list contains the main advantages of using pellets:

- Pellets burns almost without any smoke development. The dust in the flue gas is very basic.
- The ash produced is basic.
- Less carcinogens are produced in the high temperature combustion of pellets compared with unrefined fuel.
- It has a low heavy metal content.
- Only small quantities of NO_x oxides are formed.

Biomass is usually processed to pellets in plants using piston or roller presses. The raw material used may be sawdust, shavings or chips. The raw material is heated, to make the fuel dry and at the same time transforming the lignin acting like the glue that holds the raw material together, which makes it possible to shape the pellets in the desired form. To use pellets for heating, a specialised pellets burner is needed. The ash left over from the burning is rich in minerals, and can be used as a soil conditioner.

Woodchips

Woodchips.

The term woodchips refers to mechanically processed wood particles, ranging in size from 1 to 100 mm. The criteria used for woodchip quality are as follows:

- Chip size: Only the "fine" (smaller than 30 mm) and "medium" grades (below 50 mm) are suitable for small-scale installations.

- Water content: This determines the energy content of the fuel on the one hand and its storability on the other.
- Bulk density: This indicates the weight per cubic metre (bulk volume) and depends on wood type, particle shape, degree of compaction and water content.

The price of woodchips largely depends on their water content, as high water content reduces the energy content. Moisture is therefore just as important as chip size in determining the price of woodchips.

Composition of Wood

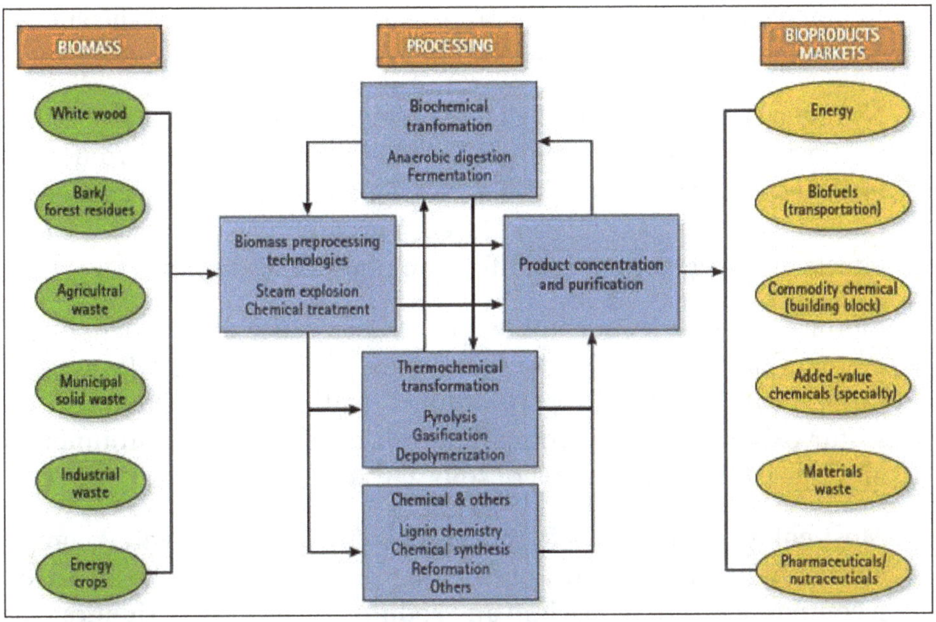

Wood is composed primarily of cellulose, hemicellulose, lignin, and mineral elements.

- Cellulose: Cellulose is the structural component of a green plant's cell walls. It is comprised of carbon, hydrogen, and oxygen in the form of starches, proteins, and sugars and is the most abundant organic material on earth. Cellulose makes up approximately 50% of woody biomass' dry weight. Cellulose from wood is isolated during the pulping process and processed to yield ethanol, cellophane, cardboard, paper, and cellulose ethers such as acetate, rayon, and nitrates.
- Hemicellulose: Hemicellulose is another component of a green plant's cell walls. Unlike cellulose, which contains only one sugar (glucose), hemicellulose can include a number of sugar monomers (xylose, hexose, mannose, and galactose, for example). Hemicellulose makes up 25 to 35% of the dry weight of woody biomass.
- Lignin: Lignin is a polymer (or glue-like substance) that holds cellulose and

hemicellulose together. It makes up approximately 15 to 25 percent of the dry weight of woody biomass and is what can make the biochemical conversion of woody biomass difficult. Lignin has not yet been used as a raw material for industrial purposes in large quantities.

- Mineral Elements: Woody biomass is composed of many elements. Principal elements include carbon, oxygen, and hydrogen. Woody biomass is also composed of mineral elements such as nitrogen, sulfur, chlorine and heavy metals and while these elements do not produce energy during combustion, they do affect the energy content of different types of woody biomass. On average, hardwoods have a higher concentration of mineral elements than softwoods. However, the presence of mineral elements in woody biomass is more dependent on the site where the particular species of tree is grown rather than the particular species itself.
  - Nitrogen is a component of all fuel systems. During the combustion process, it is oxidized into nitrogen oxide (NO_x). When emitted from combustion facilities at relatively low levels, NO_x may have a useful fertilizing effect on forests. However, as emission levels increase, NO_x produces adverse health effects and increases the acidification of water and soils.
  - Sulfur emissions from combustion of fuels cause extensive damage to ecosystems and buildings, so fossil fuels are often graded by the amount of sulfur present. As with nitrogen, sulfur is oxidized during combustion to form sulfur oxide (SO_x). This compound can have serious environmental effects and causes the acidification of soils and water.
  - Most chlorine in trees is found in the foliage as an essential component in chlorophyll. Although only present in trace amounts, its ability to form alkali compounds with potassium and sodium, resulting in oxidation and corrosion, can create serious problems for boiler equipment during combustion. Eliminating foliage from woody biomass feedstocks can reduce corrosion problems, as can cofiring biomass resources with higher sulfur content fuels such as peat or coal.
  - Heavy metals tend to vaporize during combustion. The remainder contributes to ash formation. Should levels of heavy metals be high, recycling of ash as fertilizer is restricted by environmental legislation, since the metals may leach into groundwater or be absorbed by crops.

Moisture Content

The moisture content of biomass material varies greatly and plays a large role in determining the most suitable energy conversion process. Wet conversion processes such as fermentation are often more suited to biomass with higher moisture content (e.g. corn, sugarcane, barley straw). Dry conversion processes such as pyrolysis, gasification, and combustion are more suited to biomass with lower moisture content (e.g. wheat straw, pine, switchgrass etc). Generally, wet conversion processes are used when the moisture content of the biomass requires excessive energy for drying, compared to the energy content of the end product.

Energy yields are often expressed as net caloric values. These values increase as wood moisture content is reduced. The moisture content in wood depends on a combination of climatic conditions, time of year when harvesting takes place, and the duration and method of storage. These simple formulas can calculate the moisture content of woody biomass:

Moisture content (wet basis) = (total weight of wet wood – oven dry weight) total weight of wet wood • 100

Moisture content (dry basis) = (total weight of wet wood – oven dry weight) oven dry weight • 100

Sources of Woody Biomass

Woody biomass is the solid portion of stems and branches from trees or residue products made from trees. Woody biomass can come from a variety of sources, including:

- Non-timber tree removal: Removing dead and dying trees, unwanted urban trees, or trees impeding land development.

Dense timber stand with high fuel hazard.

- Forest management harvesting: The removal of small diameter trees from overpopulated stands for wildfire hazard fuel reduction, pre-commercial thinning of timber stands, or forest health improvement.
- Timber harvesting and logging residues: Non-merchantable wood including branches, undersized trees, and non-commercial species removed during typical timber harvesting operations.
- Sawmill and other wood manufacturing residues – includes bark, undersized and defective wood pieces (e.g. lumber edging and end trimming), sawdust, and other wood waste.

Sawmill residues.

- Landfill diversion: Wood debris from tree removal and pruning, construction, demolition, discarded shipping materials (boxes, crates, pallets), and other trashed wood products.
- Chaparral management: Removal of excess woody shrubs and plants for wildfire fuel hazard reduction or other vegetation management goals.
- Dedicated forests (plantations): Fast growing trees grown specifically for biomass markets.

Dedicated poplar forest.

Biomass Utilization

Woody biomass has the potential to be used to make any of the products that are already produced or manufactured from wood. And, just like wood it benefits from the same advantages. But because woody biomass is by definition a low quality raw material there are many challenges to using the resource effectively and efficiently.

Advantages of Woody Biomass

- It is a non-food, organic material that as a feedstock will not compete with agricultural interests for growing food crops.
- It is a renewable resource.

- It has positive benefits to the environment because of the lower energy requirements to manufacture products than comparable non-wood materials (e.g. wood vs. steel).
- It can reduce wildfire hazard when it is removed from the forest or wildland urban interface.
- When burned to produce energy, emits an amount of carbon dioxide that is comparable to the amount of CO_2 released by wood during natural degradation. Since trees take in CO_2 during photosynthesis, using wood to produce energy is considered "carbon neutral."

Challenges to use Woody Biomass

- Limited availability: Difficult to obtain enough feedstock near large processing plants to take advantage of the "economies of scale."
- High cost: Associated with the harvest, transportation and storage of woody biomass.
- Low-energy density: Woody biomass has a much lower energy density (unit of energy per unit of volume) than fossil fuels.
- Technological difficulties: The recalcitrant nature of wood and its complex chemistry make it a technologically difficult feedstock to use in chemical and energy processing.
- Competition: The growth of woody biomass markets may divert higher-quality wood away from traditional wood product markets.

Woody Biomass for Commercial Forest Products

Woody biomass in tree form cannot be used to make lumber, if it could it would be classified as timber and not woody biomass. By definition, woody biomass is unsuitable for lumber. Woody biomass in tree form comes from trees that are too small to make standard lumber sizes or trees that are too low in quality (e.g. decayed) to achieve minimum lumber quality standards. Woody biomass in non-tree form is found in a size and shape unsuitable for lumber. However, many other commercial forest products can be made from woody biomass. The range of products that could be produced from woody biomass includes:

- Soil amendments: Woody biomass roughly milled into small particles can be added directly to soil as a carbon rich, degradable buffer or to other organic materials high in nitrogen content to form balanced compost.
- Landscape materials: Bark and wood chips can be processed into uniform sizes for decorative ground cover, mulch, or playground fill.

- Firewood: Woody biomass in stem form can be processed into standard firewood sizes by cutting to length and splitting.
- Posts and poles: Woody biomass in stem form can be sorted to obtain stems large enough and with enough sound wood that they can be cut to length and peeled to a uniform diameter of standard post or pole size.
- Wood fiber resource: Woody biomass treated to a strict preparation process can be cleaned up and sized appropriately to be the feedstock for composite panel production (particleboard, oriented-strand board, and fiberboard), wood-plastic lumber, or pulp for paper.
- Feedstock for bio-refineries: Contaminant free woody biomass could be the feedstock to produce numerous organic chemicals, including biofuels.

The above list is ordered in ascending value of the product produced and also in increasing cost of feedstock preparation. The use of woody biomass as a wood fiber resource for composite products or as a feedstock for bio-refineries is unlikely at present because of the high processing costs associated with procurement, transportation and preparation of the feedstock.

Woody Biomass for Bioenergy

Woody biomass and bioenergy are in some circles nearly synonymous terms. This is not surprising considering the heightened interest in alternative energy, the impacts of energy production and use on climate change, and the definite advantages related to expanding the use of woody biomass for energy. The technology exists to produce heat energy directly from woody biomass or to produce intermediate biofuels designed to be stored and transported long distances. Bioenergy categories suitable for woody biomass include the following:

- Heat Energy: Through the exothermic combustion process, wood or woody biomass, is converted into the primary products of carbon dioxide, water, inorganic ash, and various gaseous and particulate emissions while giving off about 8,000 BTU's of heat for every pound of dry wood burned.
- Electrical Energy: Coupling the combustion process with a steam boiler and using the produced steam to drive an electrical turbine is a well proven method of producing electricity from woody biomass.

Sustainability Issues

The growing diversity of uses and public expectations related to forests has led to the concept of sustainable forest management as a central purpose in managing forests. Sustainable forest management is yet to be defined, however, governments and other organizations have developed systems of Criteria and Indicators to define and describe sustainable forest management so that the range of forest activities can be assessed and their management adapted accordingly. These Criteria (values) and Indicators (measurements of values) are designed to be implemented on regional, national and international scales. Economic criteria consider levels of employment, price of wood and other forest products. Environmental criteria evaluate the health, productive capacity, biodiversity, soil, water, and carbon budgets. Social criteria take into account public participation in forest management decisions, the use of forests for spiritual and aesthetic uses, and other related activities. All combine to enable assessments of sustainability. Biomass production for energy is a product of forests, thus its benefits and impacts on forests are monitored to ensure sustainability.

Certification, an independent attestation that the products of forests are generated from sustainably managed lands, is another consideration which may soon affect biomass production. Certification is done to secure continued access to public forest lands through improved public acceptance of forest management activities. In some countries, certification that a power production system benefits the environment is being used as a marketing tool; as this gains acceptance the call for certification will become more widespread.

Economic Sustainability

The development of bioenergy markets can have many positive economic benefits including:

- Creating markets for biomass wastes.
- Improving the economic viability of thinning and harvesting.
- Promoting new crops, especially on marginal or unused agricultural land.
- Creating employment while harvesting, transport and conversion to useful energy.

In most biomass production systems the cost and value of the biomass is low, though this may discourage initiatives in biomass production, it may also stimulate the use of bioenergy if the cost is competitive with fossil fuels. Biomass production on agricultural land will become more economically attractive as agricultural policies reduce subsidies for food crops and the opportunity is enhanced for growing bioenergy crops on now-surplus land. Some farmers are already planting trees for timber or biomass, rather than leaving the land idle. The potential for mixed forestry and biomass, coupled with carbon-sequestration plantations may make tree farming an increasingly attractive economic proposition. The prospect of carbon emissions trading will further enhance this process.

Biomass from integrated harvesting operations can improve the financial return from harvesting and make operations in previously marginal areas economically viable.

Here, harvesting operations target stands where conditions are best for recovery and the scale of operations allows full utilization of equipment. In many areas there is a growing trend towards joint ventures, agro-forestry, or land leasing arrangements with farmers, so that rural communities are not displaced by expansion of planted forests. Already on some farms, willow or spruce (Picea) are being planted and economic opportunities for non-food crops is showing promise. Future energy markets are likely to be less regulated than today and the forest energy industry more subject to market forces, resulting in higher process for carbon in response to demands from international protocols and treaties. Effective policies will need to be transparent, cost-effective in achieving objectives, and `fair' as regards renewable versus non-renewable energy systems. These factors must be considered in justifying the use of biomass for energy.

Environmental Sustainability

There are environmental impacts arising from the production of biomass, as there are in managing many natural resources. Three of the more important ones are site productivity, biodiversity, and greenhouse gas balances.

Site Productivity

A commonly-expressed environmental concern about harvesting biomass for energy is that soil nutrients, organic matter and moisture-holding capacity may be depleted by intensive harvesting methods. Protection of soil relies on careful harvesting practices to reduce physical soil disturbance and compaction or removal of organic matter layers on the soil surface. Where roads and extraction tracks disturb organic layers, there is a need to manage water flows and runoff to reduce contamination of streams and water bodies by soil and silt. Soil compaction, which reduces the extent and time of root growth, can be minimized by operating when soils are dry or frozen and by avoiding repeated passes of heavy equipment. Regulations governing harvesting practices are well-established and can be readily implemented in natural or planted forests.

Nitrogen and other elements are abundant in twigs and foliage so that harvesting all above-ground biomass could theoretically remove a large proportion of nutrients. In practice this does not occur since harvesting practices remove a small portion of the branches and tops and leave sufficient biomass to conserve organic matter and nutrients. Furthermore, if nutrients are returned to the forest through ash from combustion of the residues, this ash fertilization compensates for most nutrient losses. On nutrient-poor sites the ash should be recycled once per forest rotation. Thus forest residues can be utilised much more than they are today without significant negative environmental impacts. Short-rotation crops are relatively more demanding in terms of nutrient and cultural treatments than crops from natural forests. Nevertheless, the same principles apply and productivity can be maintained by the appropriate silvicultural practices.

Science-based studies of site productivity and harvesting are now able to indicate which areas should not be harvested for biomass. Sites where nutrients are the primary limitation to tree growth should not be harvested, or harvest should be limited to removal of stemwood. Avoiding harvesting on drought-stressed sites, and limiting removals to once per rotation largely avoid the environmental impacts on harvesting.

Biodiversity

Biodiversity conservation is a central issue to forest management and is a significant public policy issue. Management of natural forests emphasises conservation of extant biodiversity by protecting critical habitat and balancing the vegetation structure, growth stages and forest ecosystems over time. In managing planted forests there is emphasis on retaining patches or riparian corridors of natural vegetation, and in some cases re-establishing native vegetation as part of plantation establishment. Natural or non-planted forests have traditionally had greater role in biodiversity conservation than plantations which are prized for production of wood fibre over other products. However, careful forest management in natural and/or planted forests can contribute to the conservation of biodiversity and to water regulation, carbon sequestration and recreational benefits.

Harvesting practices may be more intensive and so change wildlife habitats compared with conventional harvesting. Silvicultural techniques can overcome most of these effects through connection of fragmented habitats by reforestation, alteration of the size of harvested areas, elimination of pesticides, encouragement of ground vegetation, and the creation of a multi-aged, multi-species forest which provides a diverse habitat for wildlife. Experience in biomass production has shown that a normal utilisation of residues after forest operations has little negative impact on biodiversity, while the use of forest residues is environmentally beneficial because it replaces fossil fuels as an energy source. When energy crops are planted on abandoned agricultural land, species diversity may increase since diversity is low where single agricultural crops have been grown.

Short-rotation crops have much higher yields than forests so smaller areas are needed to produce biomass thus reducing the area under intensive forest management. The creation of structurally and species diverse forests also helps to reduce the impacts of insects, diseases and weeds. Similarly the artificial creation of diversity is essential when genetically modified or genetically identical species are being planted. Finally, it is through an intelligent approach to managing forests that will ensure that sustainability goals are met.

Greenhouse Gas Balance

Bioenergy systems offer significant possibilities for reducing greenhouse gas emissions when bioenergy replaces fossil fuel in energy production. The greenhouse gas balance of producing bio-energy is positive so replacement of fossil fuels with biomass both reduces emissions and enhances carbon sequestration since short-rotation crops or forests established on former agricultural land accumulate carbon in the soil (sinks). Bioenergy usually provides an irreversible mitigation effect by reducing carbon dioxide at source, but it may emit more carbon per unit of energy than fossil fuels unless biomass fuels are produced unsustainably. This is why bioenergy is so strongly promoted as a contributor to the problem of global change.

The Kyoto Protocol is stimulating policies directed towards the limitation of greenhouse gas emissions, particularly carbon dioxide (CO_2). The two main causes of rising CO_2 concentrations are the burning of fossil fuels and land-use changes, particularly deforestation. The reduction of CO_2 emissions and reversal of deforestation by afforestation are opportunities to address climate change. Substitution of biomass energy for fossil fuel-based energy will also require an effective accounting of carbon. The United Nations Framework Convention on Climate Change and the Kyoto Protocol emphasise carbon and emissions accounting in all areas of the economy.

Social Sustainability

The essence of social sustainability is how different societies benefit from biomass production. Biomass production systems require people to operate them thus creating jobs. As well, public perceptions of bioenergy systems may place different values on forests and landscapes. Most residue-harvesting operations are conducted by contractors who might supply biomass for a small district heating plant, or who collectively supply larger plants. The impact on employment is primarily in rural areas. In many countries, there remains a strong cultural tradition for the place of fuel wood biomass in the energy supply. However, as the efficiencies of scale increase, or as integrated harvesting systems are used, fewer people tend to be employed per volume of biomass harvested.

In some places cultural traditions are being revived by the increased use of woody biomass for bioenergy. In the boreal forest, many aboriginal communities have no year-round road or connections to the national electricity grid, and are dependent on diesel generators supplied by fuel flown or barged in at high cost. These communities are often surrounded by forest that could provide the necessary biomass for energy generation. This would make the community more self-sufficient, reduce costs, provide employment, and integrate well with a forest-based culture. There are examples where a shift to locally-produced bioenergy has been successful and these successes need to be encouraged.

There is a need for a radical shift and understanding by local planners and communities to take advantage of the opportunities presented by the increased use of bioenergy. Urban attitudes to biomass production are equivocal; here, we need better communication so that all citizens appreciate the use of biological systems to provide our needs sustainably. Biomass energy developments will be strongly influenced by policies and incentives to reduce greenhouse gas emissions. These developments can ensure a positive climate for investment in biomass energy to address the larger needs of society.

Biochemical Processes

Biochemical processes work thanks to the action of fungi and bacteria that grow in the biomass in specific conditions of temperature and humidity. These micro-organisms digest organic matter freeing molecules that are waste (for them) but precious for us. Not all biomasses are suitable for these processes: fungi and bacteria do not eat absolutely everything but necessitate organic matter rich in proteins and water. Seaweeds, rejects from the cultivation of potatoes and beets, food waste and animal faeces are very good. The main products that can be obtained with these methods are biogas, bioethanol, fertilizers for agriculture and heat.

Biogas is a mixture of gases, predominantly methane (50-70%) and carbon dioxide, which can be used for heating purposes or to make some particular plants allocated to the production of electrical energy work. Bioethanol is an alcohol that can be used as car fuel. It is obtained from the fermentation of the sugars of beets and sugar cane. It is a fuel of great interest because it is clean and cheap. It has been calculated that 11 million tonnes of bioethanol are produced annually, especially in U.S.A. and Brazil. Another interesting application of biomass is for heating animal farms and cultivations. The decomposition of waste products, like leaves or animal faeces produces heat that can be used to warm up greenhouses and stables.

Thermochemical Processes

It is well known that to light and feed a fire you need materials that burn and these, in technical terms, are known as fuel. Fuel alone is not enough for a fire to burn, another element is required: the combustive agent. The most common combustive agent is oxygen, which, in a combustion reaction, has the task of "oxidizing" the fuel, with the consequent release of energy in the form of heat and light. Fire therefore is simply the visible manifestation of a chemical reaction, combustion that takes place between two different substances: the fuel and the combustive agent. There are numerous combustible substances and materials. Initially man burnt wood, then coal. Today the most common fuels are the fossil fuels: petroleum, methane gas and coal or fossilized carbon.

Combustion is the most ancient method to obtain energy from the biomass. Antique fireplaces, chimneys and stoves have now been replaced by modern and efficient heaters that are able to exploit the hidden energy of wood and its derivatives. Fungi and bacteria prefer humid protein substances, but fire feeds best on dry materials that are rich in cellulose. Cellulose is a complex molecule. It is very resistant and consists of long glucose chains, the most simple of sugars. Plants are made of cellulose, so are wood, leaves, paper and cotton. There are various systems to obtain energy from wood that are classified according to the combustion temperature and the type of physical and chemical transformation that is obtained.

Firstly it must be pointed out that these systems use crushed wood. The chips of wood can be used just as they are, or compacted in small blocks or pellets. Wood pellets increase the efficiency of the heaters and leave them cleaner. Wood transformed in this way can be burnt at extremely high temperatures (about 1000°C), and it turns into a mixture of gases that can move turbines and produce electric power. When it is burnt at lower temperatures, (from 400 to 800°C) wood separates into gases, liquids and solids. The solid component, coal, can still be used as a fuel, and the liquid component, pyrolysis oil can be used as a fuel for engines or can be used as the base in the synthesis of other products.

Energy Production

Biopower

The technologies used to obtain energy (biopower) from different types of biomass are different and the resulting energy products are different too. Biopower technologies convert renewable fuels of biomass into heat and electricity by using equipment, which is similar to the one used for fossil fuels. An advantageous characteristic of biomass is its availability, as it is able to keep its energy intact until it is used. Briefly, thermo-chemical conversion processes of biomass are based on the action of heat that activates the chemical reactions needed to transform the matter into energy and can be used for those cellulose and wood residues whose C/N ratio is over 30 and the humidity content is lower than 30%. The biomass that is more suitable to be subject to thermo-chemical conversion are wood and all its derivates (sawdust, wood shavings, etc.), the most common wood-cellulose by-products (cereal straw, grapevine pruning residues, fruit trees pruning residues, etc.) and some processing waste (husks, chaff, stones, etc).

Bio-chemical Processes

Bio-chemical conversion processes allow to obtain energy throughout a chemical reaction that takes place thanks to the contribution of enzymes, mushrooms and micro-organisms, that form in the biomass under particular conditions, and are used for that biomass whose carbon/nitrogen ratio is lower than 30 and humidity at collection is higher than 30%. The following products are appropriate for chemical conversion: aquatic breeding, some cultivation by-products (leaves, sugar-beet stems, vegetables, potatoes, etc.), liquid residues of zoo-technical industry and some processing waste (residues of alcohol processing made of grains, vegetation water, etc.), and some types of urban and industrial waste water. The technologies that are currently used in the biopower sector are cofiring, pyrolisis, gasification, combustion, “small modular” systems, aerobic digestion, anaerobic digestion, and carbonization.

Gasification

At present, in terms of biomass, "gasification", which is a thermo-chemical process, is considered as one of the best and most promising technologies to produce electric energy: both as far as efficiency and environmental impact are concerned. Each plant is subdivided into three sections, where three stages of the productive process are carried out: gasification, gas turbine and thermal cycle. During gasification, the wet biomass is conveyed into a drier making excess water evaporate. After being dried, biomass proceeds to the gasifier, where it is transformed in a synthetic gas composed of molecular nitrogen (N_2), steam (H_2O), carbon monoxide (CO), carbon dioxide (CO_2), molecular hydrogen (H_2), methane (CH_4) and a small fraction of heavier hydrocarbons. Then the synthetic gas is cooled and filtered to eliminate dust, contaminants (cyanic acid, ammonia and hydrogen chloride) and organic compounds (phenols and fatty acids).

After being compressed, it is ready to operate the gas turbine where it will be burnt to heat the air to be conveyed to the thermal cycle. In the last section of the plant a boiler recovers the heat contained in the air coming from the gas turbine and produces steam for another turbine, which will operate the electric current generator.

Cofiring

In order to optimise coal plants, it is possible to use biomass as complementary to coal. This is surely one of the cheapest solutions among the energy options offered by renewable sources. The cofiring is based on the replacement of a portion of coal with biomass to be used in the same boiler located in the already existing plant. This can be done by mixing biomass with coal before the fuel is introduced into the boiler or using separate feeding systems for coal and biomass. According to the type of boiler and the feeding system used, the biomass can replace up to 15% of coal in this cofiring operation.

Pyrolysis

Pyrolysis is the thermo-chemical decomposition of organic materials that is obtained through heat application, at a temperature between 400 and 800°C, in complete absence of any oxidizing agents, or with a much reduced quantity of oxygen. The products of pyrolysis can be gaseous, liquid, solid and their proportions depend on the pyrolysis method (fast, slow or conventional) and reaction parameters. One of the main problems linked to the production of energy through pyrolysis is the quality of the products, which has not reached an adequate level in its applications, neither with gas turbines nor diesel engines. For the future, the cycles with pyrolytic oil seem to be more promising, especially for large installations, while diesel engines that use pyrolysis products, seem to be more suitable for small installations. Direct combustion generally occurs inside equipment (boilers), where the heat is exchanged between combustion gases and process fluids (water, etc).

Combustion

The combustion of products and agricultural residues has good results if substances rich in structured glucides (cellulose and lignin) and with a water content of less than 35% are used as fuels. The products that can be used are the following: wood in all its forms, cereal straws, residues of dry legumes, residues of oleaginous plants (castor-oil plant, etc.), residues of textile fibre plants (cotton, hemp, etc.), wooden residues deriving from the pruning of fruit and forest plants, residues of the agro-food industry.

"Small-Modular" Systems

These systems could potentially satisfy the energy need of 2.5 billion people that have no electric energy. This capacity is due to the fact that most of these people live in areas where large quantities of biomass can be used as fuel. A small system with a capacity of 5 megawatts could be an excellent solution in villages. Small systems have a potential market also in industrialized countries, as they could be used as a complementary energy supply. Compared to fossil fuel systems, they are a more acceptable alternative also from the environmental point of view.

Carbonization

Carbonization is a thermo-chemical process that allows transforming structured molecules of wood and cellulose products into coal (wood coal or vegetal coal). Carbonization is obtained through the elimination of water and volatile substances from the vegetal matter, due to the action of heat in charcoal pile (cone-shaped wood piles, covered with earth, with a central outlet channel (chimney), where a slow wood combustion takes place, turning the wood into coal). Carbonization occurs outdoors, or in long and curved-neck containers, with a flask shape, that offer a higher coal yield.

Anaerobic Digestion

It is a biochemical conversion process that occurs in the absence of oxygen and consists of the demolition, by microorganisms, of complex organic substances (lipids, proteins, glucides) contained in vegetal and animal by-products, which produces a gas (biogas) made of methane (50-70%) and the rest is mainly CO_2 and has an average calorific value of 23,000 kilojoules per cubic metre. The resulting biogas is collected, dried, compressed and stored, and can be used as a fuel to feed gas boilers and produce heat (also coupled with turbines for the production of electric energy), or to feed combined-cycle plants, or internal combustion motors (boat engines with a low number of turns are suitable for this). At the end of the effluent fermentation process the main nutritional elements that were already present in the raw material are kept intact (nitrogen, phosphorus, potassium), by favouring the mineralization of organic nitrogen. The effluent results to be an excellent fertilizer. Anaerobic digestion plants can be fed through residues with high humidity content, like animal faeces, civil waste (waste water), food waste and the organic fraction of urban solid waste. However, also in those dumps that are suitably equipped for the collection of biogas, only 40% of the gas produced can be collected, while the remaining part is dispersed into the atmosphere. As the methane, that largely composes biogas, is a greenhouse gas with an effect that is twenty times as high as CO_2 emissions of biogas into the atmosphere are not desirable. When the decomposition of organic waste is obtained through anaerobic digestion of (closed) adequate plant digestors, almost all the gas is collected and used as a fuel. The recovery of the biogas from dumps is a system that has been experimentally adopted in various countries. In Sweden, there are biogas refueling stations that supply methane vehicles.

Aerobic Digestion

It is the metabolization of organic substances through micro-organisms, whose development depends on the presence of oxygen. These bacteria convert complex substances into simple ones, releasing CO_2 and H_2O and highly warming up the substrate, according to their methabolic activity. The resulting heat can then be trasferred outside, through fluid heat exchangers. In Europe the aerobic digestion system is used to treat waste waters. More recently, this technology has spread to Canada and the United States.

Effect of Plant Characteristics

Biomass contains varying amounts of cellulose, hemicellulose, lignin and a small amount of other extractives. Woody plant species are typically characterised by slow growth and are composed of tightly bound fibres, giving a hard external surface, while herbaceous plants are usually perennial, with more loosely bound fibres, indicating a lower proportion of lignin, which binds together the cellulosic fibres: both materials are examples of polysaccharides; long-chain natural polymers. The relative proportions of cellulose and lignin is one of the determining factors in identifying the suitability of plant species for subsequent processing as energy crops. Cellulose is a glucose polymer, consisting of linear chains of (1,4)-D-glucopyranose units, in which the units are linked 1–4 in the b-configuration, with an average molecular weight of around 100,000.

Hemicellulose is a mixture of polysaccharides, composed almost entirely of sugars such as glucose, mannose, xylose and arabinose and methylglucuronic and galaturonic acids, with an average molecular weight of <30,000. In contrast to cellulose, hemicellulose is a heterogeneous branched polysaccharide that binds tightly, but non-covalently, to the surface of each cellulose microfibril. Hemicellulose differs from cellulose, in consisting primarily of xylose and other five-carbon monosaccharides.

Lignin can be regarded as a group of amorphous, high molecular-weight, chemically related compounds. The building blocks of lignin are believed to be a three carbon chain attached to rings of six carbon atoms, called phenyl-propanes. These may have zero, one or two methoxyl groups attached to the rings, giving rise to three structures, termed I, II and III, respectively. The proportions of each structure depend on the source of the polymer i.e. structure I is found in plants such as grasses; structure II in the wood of conifers; while structure III is found in deciduous wood. Cellulose is generally the largest fraction, representing about 40–50% of the biomass by weight; the hemicellulose portion represents 20–40% of the material by weight.

Both woody and herbaceous plant species have specific growing conditions, based on the soil type, soil moisture, nutrient balances and sunlight, which will determine their suitability and productive growth rates for specific, geographic locations. Many types of perennial grasses, such as sugarcane and cereals like wheat and maize, have widely different yields, depending on the growing conditions: thus wheat can be grown in both hot and temperate climates with a wide range of rainfall, whereas sugarcane can be grown successfully only in warm, moist climatic conditions.

Photosynthesis

Photosynthesis is the process by which chlorophyll containing organisms – green plants, algae, and some bacteria – capture energy in the form of light and convert it to chemical energy. Virtually all the energy available for life in the Earth's biosphere, the zone in which life can exist, is made available through photosynthesis. A generalised, unbalanced, chemical equation for photosynthesis is,

$$CO_2 + 2\,H_2A + \text{light energy} = \left(CH_2O\right) + H_2O + A_2$$

The formula H_2A represents a compound that can be oxidised, i.e. from which electrons can be removed and CH_2 is a general formula for the carbohydrates incorporated by the growing organism. In the vast majority of photosynthetic organisms – that is, algae and green plants – H_2A is water (H_2O) and A_2 is oxygen (O_2); in some photosynthetic bacteria however, H_2A is hydrogen sulphide (H_2S). Photosynthesis involving water is the most important and best understood and therefore will be discussed in more detail.

Photosynthesis consists of two stages: a series of light-dependent reactions that are temperature-independent and a series of temperature-dependent reactions that are light independent. The rate of the first series, called the light reaction, can be increased by increasing light intensity (within certain limits) but not by increasing temperature. In the second series, called the dark reaction, the rate can be increased by increasing temperature (within certain limits) but not by increasing light intensity.

Another differentiator amongst plant species is the type of photosynthetic pathway utilised by the plant. Most plants utilise the C_3 photosynthesis route, the C_3 determining the mass of carbon contained in the plant material. Another photosynthesis pathway is represented by C_4 plants, which accumulate a significantly greater dry mass of carbon than do C_3 plants, giving a biomass with increased potential for energy conversion. Examples of C_3 species are poplar, willow, wheat and most other cereal crops, while the perennial grass, Miscanthus, sweet sorghum, maize and artichoke, all use the C_4 route.

Traditionally, the biochemical conversion of biomass into liquids, such as ethanol, has been undertaken using sugar/starch feedstocks, such as cereals. The conversion of cellulose into glucose via acid/enzymatic hydrolysis and the subsequent conversion of glucose into alcohol, using fermentation, is more easily undertaken with highcellulose content biomass, than with lignin-rich biomass i.e. woody species, which is not so easily converted.

Plant Species

The choice of plant species depends upon the end-use, bio-conversion option of interest e.g. combustion, gasification, pyrolysis, fermentation or mechanical extraction of oils. Some plant species are amenable to nearly all of the potential conversion technologies: e.g. oil seed rape can be processed via combustion, gasification, pyrolysis or mechanical extraction, while others such as wood and cereal crops are suitable for combustion, gasification, pyrolysis and fermentation.

It is important to note that, while particular plant species may have specific benefits for subsequent processing technologies, the amount of energy potentially available from a given biomass source is the same, irrespective of the conversion technology used. What will vary between conversion technologies is the actual amount of energy recovered from the biomass source and the form of that energy.

The attention paid to particular woody/herbaceous plant species varies around the world, taking account of the soil and climatic factors that affect growth. In the context of northern Europe, much attention has been focused on the C_3 woody species, especially those grown as short rotation coppice (SRC) e.g. willow and poplar and forestry residues. Herbaceous C_3 species, such as cereals, are of less interest in the UK due to limited interest in the bio-conversion of carbohydrate-rich biomass to alcohol (ethanol) fuels, which is not the case in the USA and some parts of Europe. Similarly oil seed rape, for producing bio-diesel fuel, is of wide interest in Europe but not so in the UK. In tropical climates where sugarcane can be grown, Brazil has been one of the pioneers of large-scale fuel–alcohol production derived from sugarcane.

Of the herbaceous plant species, the perennial C4 grass, Miscanthus, has created considerable interest. Miscanthus has been identified as the ideal fuel crop, providing an annual crop, being easy to grow and harvest and when harvested dry, gives a high dry-matter yield. Light, arable soils give a good yield, provided there is adequate rainfall, which would not be a problem in the UK: dark-coloured soils produce a better yield than light-coloured soils and the plant is pH tolerant. In the UK, yields would overall be greater in the wetter West than the drier East of the country. This thin-stemmed grass has been evaluated as a bio-energy crop in Europe for over 10 years, having been grown experimentally in over 10 countries. The combination of an annual harvest, low inherent mineral content and a good energy yield/ha gives the plant considerable potential as a bio-energy crop in the UK.

Miscanthus is currently propagated as rhizomes, planted in double rows about 75 cm apart, with 175 cm gaps between double rows. Weed control is vital for crop establishment but the plant faces negligible disease problems within Europe. Yields of 3–10 years old plantations grown in Germany and Denmark are 13–30 t/ha: if a yield of 20 t/ha could be achieved, this would give a gross energy yield equivalent to 7 t/ha of oil over the life of each crop. On the basis of a 20 t/ha yield of dry matter, the value as an energy crop would be about £620=ha, compared with wheat at £400 ha; this equates to a value of one third that of oil and about half that of coal. Overall, the production costs of Miscanthus, based on a 20 t/ha yield, compare favourably with wood chips from SRC forestry and with whole crop cereals.

Biofuels

Any hydrocarbon fuel that is produced from an organic matter (living or once living material) in a short period of time (days, weeks, or even months) is considered a biofuel. Biofuels may be solid, liquid or gaseous in nature.

- Solid: Wood, dried plant material, and manure.
- Liquid: Bioethanol and Biodiesel.
- Gaseous: Biogas.

These can be used to replace or can be used in addition to diesel, petrol or other fossil fuels for transport, stationary, portable and other applications. Also, they can be used to generate heat and electricity. Some of the main reasons for shifting to biofuels are the rising prices of oil, emission of the greenhouse gases from fossil fuels and the interest for obtaining fuel from agricultural crops for the benefit of farmers.

Categories of Biofuels

- First generation biofuels:
    - These are made from food sources such as sugar, starch, vegetable oil, or animal fats using conventional technology.
    - Common first-generation biofuels include Bioalcohols, Biodiesel, Vegetable oil, Bioethers, Biogas.
    - Though the process of conversion is easy, but use of food sources in the production of biofuels creates an imbalance in food economy, leading to increased food prices and hunger.

- Second generation biofuels:
    - These are produced from non-food crops or portions of food crops that are not edible and considered as wastes, e.g. stems, husks, wood chips, and fruit skins and peeling.
    - Thermochemical reactions or biochemical conversion process is used for producing such fuels. Examples include cellulose ethanol, biodiesel.
    - Though these fuels do not affect food economy, their production is quite complicated.
    - Also, it is reported that these biofuels emit less greenhouse gases when compared to first generation biofuels.
- Third generation biofuels:
    - These are produced from micro-organisms like algae. Example - Butanol.
    - Micro-organisms like algae can be grown using land and water unsuitable for food production, therefore reducing the strain on already depleted water sources.
    - One disadvantage is that fertilizers used in the production of such crops lead to environment pollution.
- Fourth generation biofuels:
    - In the production of these fuels, crops that are genetically engineered to take in high amounts of carbon are grown and harvested as biomass.
    - The crops are then converted into fuel using second generation techniques.
    - The fuel is pre-combusted and the carbon is captured. Then the carbon is geo-sequestered, meaning that the carbon is stored in depleted oil or gas fields or in unmineable coal seams.
    - Some of these fuels are considered as carbon negative as their production pulls out carbon from environment.

Table: Major Types of Biofuels.

Major Types of Biofuels	
Bioethanol	• It is derived from corn and sugarcane using fermentation process. • A litre of ethanol contains approximately two thirds of the energy provided by a litre of petrol. • When mixed with petrol, it improves the combustion performance and lowers the emissions of carbon monoxide and sulphur oxide.

Biodiesel	• It is derived from vegetable oils like soybean oil or palm oil, vegetable waste oils, and animal fats by a biochemical process called "Transesterification." • It produces very less or no amount of harmful gases as compared to diesel. • It can be used as an alternative for the conventional diesel fuel.
Biogas	• It is produced by anaerobic decomposition of organic matter like sewage from animals and humans. • Major proportion of biogas is methane and carbon dioxide, though it also has small proportions of hydrogen sulfide, hydrogen, carbon monoxide and siloxanes. • It is commonly used for heating, electricity and for automobiles.
Biobutanol	• It is produced in the same way as bioethanol i.e. through the fermentation of starch. • The energy content in butanol is the highest among the other gasoline alternatives. It can be added to diesel to reduce emissions. • It serves as a solvent in textile industry and is also used as a base in perfumes.
Biohydrogen	• Biohydrogen, like biogas, can be produced using a number of processes such as pyrolysis, gasification or biological fermentation. • It can be the perfect alternative for fossil fuel.

Advantages of Biofuels

- Availability: Biofuels are produced from biomass and thus are renewable.
- Source material: Whereas oil is a limited resource that comes from specific materials, biofuels can be manufactured from a wide range of materials including crop waste, manure, and other byproducts.
- Environment Pollution: Biofuels do not release as much carbon as fossil fuels do but fertilizers that are used in the growing bio fuels lead to greenhouse emissions. Also, biofuels can help in managing the municipal solid wastes i.e. the waste can be converted into fuel.
- Security: Biofuels can be produced locally, which decreases the nation's dependence upon foreign energy. By reducing dependence on foreign fuel sources, countries can protect the integrity of their energy resources and make them safe from outside influences.
- Economic stimulation: Because biofuels are produced locally, biofuel manufacturing plants can employ hundreds or thousands of workers, creating new jobs in rural areas. Biofuel production will also increase the demand for suitable biofuel crops, providing economic stimulation to the agriculture industry.

Disadvantages of Biofuels

- Efficiency: Fossil Fuels produce more energy than some of the biofuels. E.g. 1 gallon of ethanol produces less energy as compared to 1 gallon of gasoline (a fossil fuel).
- Cost: Pumping fossil fuels from the ground is a difficult and expensive process leading to high costs. Production of biofuels require land, this impacts cost of biofuels as well as that of food crops. Also, though growing engineered biofuel crops can benefit farmers commercially but the excess number of such crops can also lead to loss of biodiversity.
- Food shortages: There is concern that using valuable cropland to grow fuel crops could have an impact on the cost of food and could possibly lead to food shortages.
- Water use: Massive quantities of water are required for proper irrigation of biofuel crops as well as to manufacture the fuel, which could strain local and regional water resources.

International Initiatives on Sustainable Biofuels

Production of sustainable biofuels is required to reduce their impact on environment and economy.

- Roundtable on Sustainable Biomaterials (RSB):
    - It is an international initiative which brings together farmers, companies, governments, non-governmental organizations, and scientists who are interested in the sustainability of biofuel production and distribution.
    - In April 2011, it launched a set of comprehensive sustainability criteria - the "RSB Certification System." Biofuels producers that meet these criteria are able to show buyers and regulators that their product has been obtained without harming the environment or violating human rights.
- Sustainable Biofuels Consensus:
    - It is an international initiative which calls upon governments, the private sector, and other stakeholders to take decisive action to ensure the sustainable trade, production, and use of biofuels.
- Bonsucro:
    - It is an international not for-profit, multi-stakeholder organization established in 2008 to promote sustainable sugar cane.

- Its stated aim is to reduce 'the environmental and social impacts of sugar-cane production while recognising the need for economic viability'.
- It does that through setting sustainability standards and certifying sugar cane products including ethanol, sugar and molasses.

Bioenergy

Bioenergy refers to electricity and gas that is generated from organic matter, known as biomass. This can be anything from plants and timber to agricultural and food waste and even sewage. The term bioenergy also covers transport fuels produced from organic matter.

How does Biomass Generate Energy?

When biomass is used as an energy source, it's referred to as 'feedstock'. Feedstocks can be grown specifically for their energy content (an energy crop), or they can be made up of waste products from industries such as agriculture, food processing or timber production. Dry, combustible feedstocks such as wood pellets are burnt in boilers or furnaces. This in turn boils water and creates steam, which drives a turbine to generate electricity.

Wet feedstocks, like food waste for example, are put into sealed tanks where they rot and produce methane gas (also called biogas). The gas can be captured and burnt to generate electricity. Or it can be injected into the national gas grid and be used for cooking and heating. Bioenergy is a very flexible energy source. It can be turned up and down quickly to meet demand, making it a great backup for weather-dependent renewable technologies such as wind and solar.

Is Bioenergy Environmentally Friendly and Sustainable?

Burning biomass does release carbon dioxide. But, because it releases the same amount of carbon that the organic matter used to produce it absorbed while it grew, it doesn't break the carbon balance of the atmosphere. In comparison, burning fossil fuels releases carbon dioxide that has been locked away for millions of years, from a time when the earth's atmosphere was very different. This adds more carbon dioxide into our current atmosphere, breaking the carbon balance. The overall sustainability and environmental benefits of bioenergy can depend on whether waste feedstocks or energy crops are being used.

Waste Feedstocks

Waste biomass gives off gases naturally when it rots. If this happens in a place where there's no oxygen, such as food waste buried deep within landfill, it can generate methane which is a much stronger greenhouse gas than carbon dioxide. Instead of allowing methane to vent into the atmosphere, breaking it down in a sealed tank allows it to be captured and burnt. Burning methane leaves you with carbon dioxide and water, which are better for the environment.

Energy crops are grown specifically for generating energy. So, unlike capturing methane from waste, there isn't an argument that burning them reduces greenhouse gases which would have been given off anyway. However, energy crops can still be low carbon if they are managed sustainably. For example, when energy crops are burnt, equivalent crops should be planted that will absorb the same amount of carbon that was released by burning.

IEA Bioenergy TCP

The task of IEA Bioenergy is to contribute to the elimination of environmental, institutional, technological and financial barriers to the use of bioenergy technologies in the future. The focus is on the initiation, coordination and promotion of research, development and demonstration projects through international cooperation and the targeted exchange of information between experts from research, industry and politics in the participating countries.

This strategy is intended to help drive the development and commercialization of environmentally friendly, efficient and cost-effective bioenergy technologies. The cooperation takes the form of thematic networks ("Tasks") and is led by an Executive Committee to which the participating countries send a representative.

At present, 24 countries from Europe and overseas and the European Commission are participating in IEA Bioenergy. The cooperation thus enables a worldwide transfer of information and the coordination of national programs and research in the field of bioenergy use. In IEA Bioenergy, national experts from research, politics and industry work closely together with experts from other countries. Cooperation with IEA Bioenergy supports and promotes:

- The research and development work through the international exchange of knowledge.
- The international dissemination of the results of the papers.
- Networking with relevant stakeholders.
- The initiation of international R&D projects and scientific exchange programs.
- The establishment of contacts between companies and international companies with the aim of cooperation.

In addition, active participation in IEA Bioenergy represents a point of contact for all national task participants in the Implementing Agreement and ensures smooth coordination within national tasks. Information about the work in the tasks is actively disseminated: on the one hand through lectures at national and international events as well as articles in specialized media, on the other hand reports and information are available from the representatives in the respective tasks and via the BMK. With this active information policy, institutions that are not directly involved in the tasks should be given access to the current state of knowledge at the international level.

The current Strategic Plan 2015 - 2020 builds on previous strategies and consistently pursues them for the next 5 years, taking into account current developments. One goal is to support the commercialization of environmentally compatible, socially accepted and competitive bioenergy systems, thus contributing to the dissemination of bioenergy and bioenergy technologies in all energy markets.

IEA Bioenergy as a platform for international research and development cooperation aims to overcome institutional, technological and societal barriers that hamper the medium to long-term development of bioenergy. In addition to the topics of biomass resources, supply systems, conversion, end products and cross-cutting issues along the value chain, the current Strategic Plan includes the new topics of production and conversion of microalgae and macroalgae as well as artificial photosynthesis. Co-operation with relevant Technology Collaboration Programmes of the IEA and with other international organizations is also to be intensified.

Biomass Composites

One of the greatest environmental challenges today is to find novel ways to utilize waste and residues originating from industrial or agricultural processes. Besides raw materials for biochar, activated carbon, biogas, and biofuel production, huge amount of biomass is still underutilized. The possibility to utilize organic wastes and biomass residues as additives or reinforcements in polymer composites has attracted substantial interest especially during the present decade.

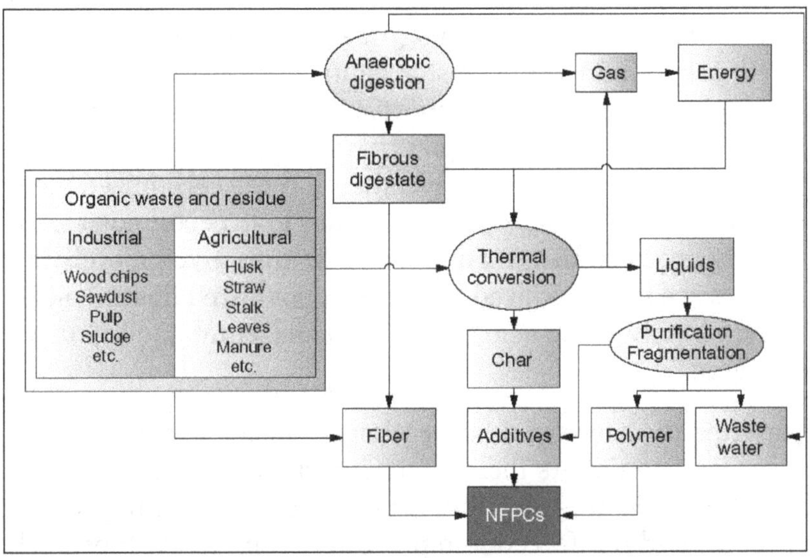

Possible ways for the utilization of organic waste and residues in NFPCs.

Figure presents some possible ways for the utilization of organic waste and residues in natural fiber reinforced polymer composites (NFPCs). Biomass or NFPCs offer multiple advantages over traditional thermoplastic composite materials with conventional reinforcements, such as E-glass, Kevlar, and carbon fiber, for example, reductions in processing temperature, cycle time, and density. Researchers reported the development of polypropylene hybrid composites by using coir fibers and coir shell particles at a total weight content of 20%. Recently, green composites were prepared with polypropylene matrix and 20wt.% spent coffee ground (SCG) powder for uses as a wood plastic composite (WPC), developed by twin-screw corotating extruder and injection molding. In other interesting study, rice straws/coir fibers are reinforced in phenol formaldehyde resin to fabricate hybrid particleboard composites with high-performance, cost-effective, and environment-friendly approach. Fabrication of polypropylene composites filled with mixtures of household waste of mate tea and *Eucalyptus benthamii* particles is receiving higher interest as a good alternative for the use of renewable materials in the production of polymeric composites; besides this, it also increases the economic value chain of mate tea. In another study, crop stalks as precursors are used to prepare two-dimensional hierarchical porous carbon composites for high-performance batteries, anode and cathode, in which the porous carbon nanosheets serve as host carbon matrixes for functional nanoparticles.

Recently, olive stone flour (OSF) from the solid residue of the olive oil extraction is used as reinforcement for the preparation of a cost-effective polypropylene-based composite. In another study, underutilized celery root fibers (*Apium graveolens*), pomace (*Vitis vinifera Chambourcin*), poplar (*Populus alba sp.*) seed hair fibers, and Asclepias syriaca fibers were embedded into a PLA matrix to fabricate green composites. Researchers also reported the utilization of underutilized nontoxic tamarind fruit fiber (TFF) as reinforcement in unsaturated polyester matrix to optimize the mechanical properties. Researchers also investigate the morphology and mechanical properties of carbonized waste maize stalk ash reinforced polyester composites for producing an ecofriendly composite material having enhanced properties for engineering applications. The tensile strength, tensile modulus, and compressive strength value increase as the carbonized maize stalk ash content increases, revealing that the carbonized maize stalk ash is a potential filler to improve the strength of polymer matrix composites for use in automobile and building applications. In another work, attractive synergistic plasticization and coupling effects are being realized by adding epoxidized linseed oil (ELO) to green composites based on poly(lactic acid) matrix and hazelnut shell flour (HSF) without further modifications.

Researchers reported the successful fabrication of plaster-reinforced Tunisian natural fiber composites. Tunisian fibers include alfa, rush, date palm stipes, date palm, and dwarf palm leaflet. Researchers also utilized the napier grass fibers extracted through water retting process and reinforced it in polyester resin. In another work, composites of neat recycled high-density polyethylene (rHDPE) and durian peel fiber were prepared by successive twin-screw extrusion blending. Researchers reported the development of high-performance biocomposites with the reinforcements of eggshell powder from sustainable resources in PLLA and its practical application as PLLA-based bioplastic. High-density polyethylene (HDPE) composites reinforced with treated biofiller from argan nut shell (ANS) at various filler contents fabricated through extrusion and injection molding processes are receiving great attention and offer an ecological alternative to upgrade the valorization of abundant and unexploited Moroccan resources. Walnut shell as a reinforcing filler in polypropylene with and without maleic anhydride grafted at 40%, 50%, and 60% (weight) contents was fabricated by injection molding.

The developed composite shows significant competitive advantages for thermoplastic composites for outdoor applications, requiring a high dimensional stability, such as deck flooring or siding. Recently, the performance of recycled polypropylene (RPP)/ peanut shell powder (PSP) composites is investigated under the effect of chemical modification of peanut shell powder (PSP) using polyvinyl alcohol (PVOH). In another work, durian skin fiber reinforced polypropylene was fabricated using extrusion and injection molding processes for robust application.

Researchers reported the feasibility of utilizing grain by-products such as wheat husk and rye husk as alternative fillers for soft wood fiber as reinforcement in composite material. In another interesting study, paddy straw powder (PSP) shows potential as a new reinforcement residual plant fiber from agricultural waste streams. Literature also reported the suitability of using hemp straw, hemp pomace as a waste biomass, and cellulose microfibers isolated from hemp straw as reinforcement in thermoplastic composites. Researchers stated the potential reinforcements of *Pennisetum purpureum* stem for the fabrication of polymer composites.

Applications of Biomass based Polymer Composites

Agricultural- and forestry-based biomass polymer composites provide a more sustainable, environmentally friendly and inexpensive structural material alternative over traditional polymer-reinforced glass or carbon fiber composite materials. These composites have a great potential for certain structural applications in building products and automotive components. Several attempts and research works have been reported for the extensive applications of biomass waste-based polymer composites. Literature revealed the development of biodegradable bark cloth reinforced green epoxy composites and its application as an alternative material for making automotive interior instrument panels. The developed biocomposites possess tensile strength of 33MPa, higher than the 25MPa threshold strength needed for car instrument or dashboard panels and flexural strength of 207MPa. It also exhibits glass transition temperature in the range of 163–185°C. The schemes of fabricating bark cloth reinforced green epoxy composites for interior automotive panels are displayed in figure.

Scheme of processing of bark cloth biocomposites for automotive applications.

In another interesting study, pomace fibers/PLA-based composites exhibited excellent antimicrobial activity against Escherichia coli and Staphylococcus aureus even after

24 h incubation time. Thermal stability and specific strength of tamarind fruit fiber (TFF)-reinforced unsaturated polyester polymer composite materials signify its applicability for lightweight automotive, aerospace, and marine applications alternative to the hazardous E-glass fibers. Recently, researchers reported a comparative evaluation between E-glass and hemp fiber composites based on a preliminary project of a helicopter device, to investigate the use of natural fiber composites for semistructural applications, such as electronic racks application and for rotorcraft interiors.

References

- Biomass-Overview, stevehar: faculty.washington.edu, Retrieved 08, March 2020
- Woody-biomass-properties: extension.org, Retrieved 21, July 2020
- Biofuels-1, to-the-points: drishtiias.com, Retrieved 16, January 2020
- Bioenergy, technologyprogrammes: nachhaltigwirtschaften.at, Retrieved 02, April 2020
- Introduction-to-biomass-and-its-composites-317507651: researchgate.net, Retrieved 19, June 2020

Chapter 2

Biomass Conversion and Heating Systems

Biomass conversion involves the usage of heat or thermal energy for the conversion of biomass energy into another form of chemical energy. Biomass heating systems replace fossil fuels for in different types of heating systems such as radiators and warm air heating systems. The topics elaborated in this chapter will help in gaining a better perspective about biomass conversion and biomass heating systems.

Thermal Conversion of Biomass

Currently utilized methods of thermal biomass conversion include combustion, gasification, biocarbonisation, torrefaction, dry distillation and pyrolysis. In relation to waste biomass, these processes are sometimes called energy recycling (direct combustion, combustion after gasification), material recycling (gasification, torrefaction, dry distillation and pyrolysis) and chemical recycling (producing biocarbonate, gases (CO, H_2, CH_4 and other), liquid complex products (tar, turpentine, pyrenes, phenols, etc.) and simple chemical compounds (pyrene, toluene, methanol, limonene, etc).

Methods of Thermal Biomass Conversion

Depending on the temperature and amount of oxygen involved in the thermal decomposition of biomass, the following versions of the process are possible: drying, torrefaction, carbonization, pyrolysis, gasification and combustion, which can be characterized as follows:

- Hydrothermal carbonization (HTC) (180–250 °C) + hot pressurized water.

  $CH_{1.4}O_{0.6} + H_2O \rightarrow$ hydrocarbonsolid fuel fulfilling the European standard (EN 14961 - 6)) + gaseous and liquid residues.

- Torrefaction:

  $$\text{Biomass } (200-300°C) \rightarrow 30\% \left(\text{gases} + \text{volatiles}\right) + 70\% \text{ torrefied biomass}$$

  $$CH_{1.4}O_{0.6} \rightarrow C_{0.7}H_{1.0}O_{0.4}(\text{solid}) + C_{0.3}H_{0.4}O_{0.2}(\text{gases})$$

- Slow pyrolysis (carbonization):

  $$\text{Biomass} + O_2\,(\text{a small amount at the beginning})(280-550°C) \rightarrow \text{biochar}$$

  $$CH_{1.4}O_{0.6} \rightarrow C + \text{tar} + 0.6\,H_2O + 0.1\,H_2 \text{and other gas products}$$

- Fast pyrolysis: $\text{Biomass}(500-800\,°C) \rightarrow \text{biochar} + \text{oil} + \text{flammable gas}$

  $$CH_{1.4}O_{0.6} \rightarrow \text{about 200 different volatile compounds}(\text{oil, tar}) + C + CO + H_2O + H_2 + \text{and other flammable gases}$$

- Gasification: $\text{Biomass} + O_2\,(\text{limited amount})(800-1000°C) \rightarrow \text{flammable gas},$

  $$CH_{1.4}O_{0.6} + 0.2\,O_2 \rightarrow CO + 0.7\,H_2\,(\text{theoretically})$$

  $$CH_{1.4}O_{0.6} + 0.4\,O_2 \rightarrow 0.7\,CO + 0.3\,CO_2 + 0.6\,H_2 + 0.1\,H_2O(\text{technically})$$

- Supercritical steam gasification: $\text{Biomass} + H_2O \rightarrow H_2 + CO_2,$

  $$CH_{1.4}O_{0.6} + 1.4\,H_2O \rightarrow CO_2 + 2.1\,H_2$$

- High temperature steam gasification: $\text{Biomass} + H_2O(>1000°C) \rightarrow CO_2 + H_2$

  $$C_mH_n + 2m\,H_2O \rightarrow m\,CO_2 + (2m + n/2)H_2$$

- Combustion:

  $$\text{Biomass} + O_2\,(\text{stoichiometric amount}) \rightarrow \text{thermal energy + flue gas}$$

  $$CH_{1.4}O_{0.6} + 1.05\,O_2 + (3.95\,N_2) \rightarrow CO_2 + 0.7\,H_2O + (3.95\,N_2)$$

The above division does not include catalytic pyrolysis processes. Depending on the coefficient of the amount of oxygen consumed in the process of thermal decomposition of biomass ϕ, which is the ratio of the mass of oxygen supplied in the process to the amount of biomass, the process varies from pyrolysis, through gasification, to complete combustion of biomass. Complete combustion of biomass occurs for stoichiometric amounts of oxygen or air, which correspond to the coefficients: $f_c = 1.476$, for pure oxygen and $f_c = 6.36$ for air. Figure presents the impact of the relative coefficient of the amount of air consumed, which is the ratio of the actual coefficient of the amount of oxygen consumed to the value of this coefficient at total combustion $\Phi = \phi/f_c$, on the course of thermal decomposition of biomass at atmospheric pressure (p = 1 atm.)

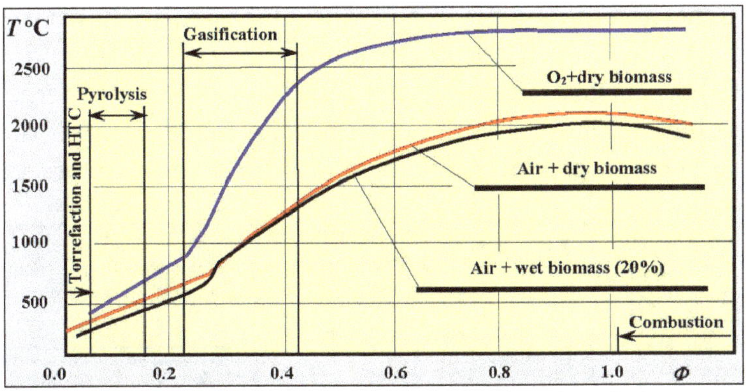

Impact of the amount of oxygen or air consumed, expressed by the relative oxygen factor Φ, on the thermal biomass degradation process, including torrefaction, hydrothermal carbonization (HTC), pyrolysis, gasification and combustion processes.

From an ecological standpoint, the least favorable process is the combustion of biomass, and therefore only devalued waste biomass originating, e.g., from municipal waste (RDF—Refuse Derived Fuel) or from sewage treatment plants (screenings, activated sludge) should be utilized using this method. All other types of biomass, both full-valued and waste, should be subjected to initial thermal conversion (torrefaction, pyrolysis or gasification) before final combustion. As a result of this material or chemical recycling, many valuable products, semi-products and chemical compounds are recovered, otherwise obtained through industrial syntheses, which are time- and work-consuming, more expensive and negatively impact the environment.

Mechanisms of Dry Thermal Decomposition of Biomass

Cellulose is the main component of biomass for energy utilization. Initially, it was believed that its thermal decomposition depends only on the temperature, proceeds in two stages: initial decomposition at temperatures below 300 °C, and further degradation and depolymerization at temperatures above 300 °C.

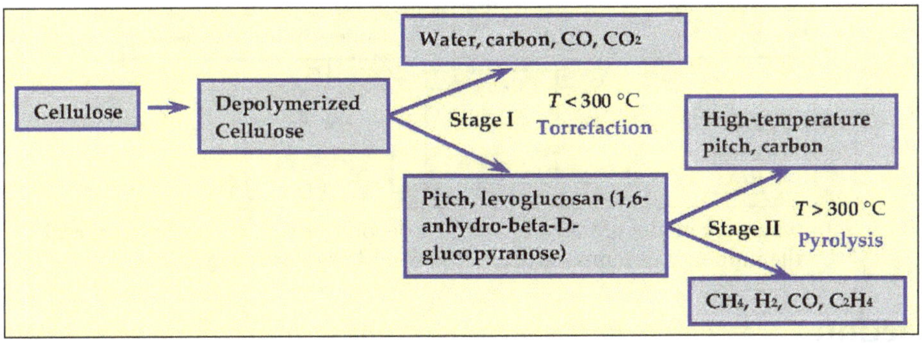

Stages of thermal decomposition of the main component of cellulose biomass.

In Stage I, cellulose can decompose according to two competing mechanisms: into biochar (in case of decomposition of wood into charcoal), CO_2 and H_2O or into tar consisting of cellulose fragments and monomers, mainly levoglucosan, as shown in above figure. As temperature increases, in Stage II, further decomposition occurs, resulting in liquid (high temperature pitch) or gaseous products. Not only the value of temperature but also the rate of its increase have an impact on the course of thermal decomposition, as well as the composition of the process' products. The results of these tests are presented in figure.

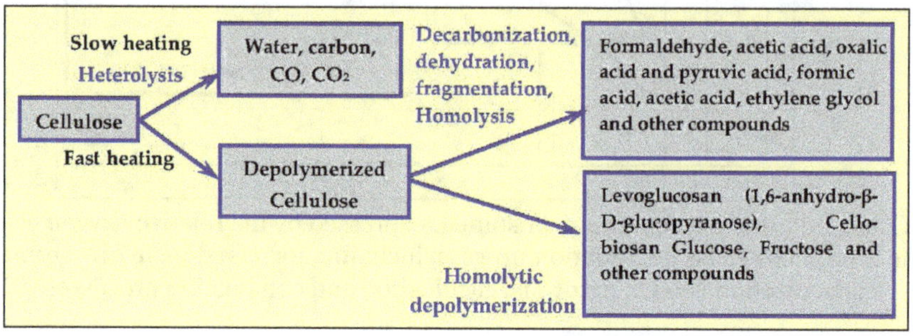

Influence of temperature increase rate on the composition of cellulose pyrolysis products.

A complete picture of thermal decomposition of biomass was obtained after considering the decomposition mechanisms—not only of cellulose, but also of other biomass components: hemicellulose and lignin, as well as the impact of pressure on the process. The first holistic thermal biomass decomposition model, taking temperature, rate of decomposition and pressure into consideration, was developed as a result of the consensus of a specialist work group. The comprehensive model proposed has been schematically presented in figure below. The impact of considered parameters: temperature T, heating rate τ and pressure p on the directions of thermal biomass decomposition is marked with arrow directions, which determine either increase or decrease in the value of these parameters in relation to reference values of T_s, τ_s and p_s.

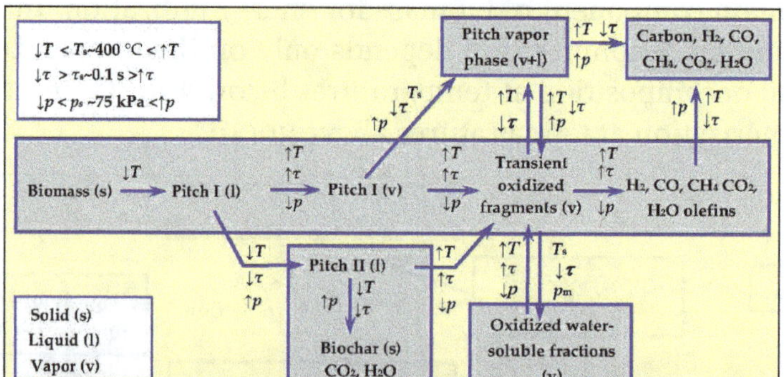

A comprehensive model of thermal biomass decomposition taking into account the influence of temperature, velocity and pressure of pyrolysis.

Torrefaction

Due to large volume and low weight, proper transport of biomass across longer distances is unprofitable. Therefore, energy plantations and other biomass sources must be located within a radius of up to 100 km from the energy processing plant. If, however, this is not possible, biochar, which has a much higher concentration of energy, can be delivered to heating or cogeneration plants instead of biomass. The density of biochar is not as high as the density of coal, but the range of its profitable transport to heat and power plants is within a few hundred kilometers. In addition, biochar can be burned in typical coal-fired boilers, in contrast to biomass, which requires special boilers. It can also be co-combusted with coal in coal-fired boilers, but the share of biochar cannot exceed approximately 15%.

Aside from the possibility of co-firing biochar in any solid fuel boiler without the need for alteration and modernization, another advantage of this fuel is its eco-friendliness resulting from a lack of sulfur, the possibility of using ash in agriculture and the fact that, unlike coal, oil or natural gas, biochar is a renewable energy source. As an effect of torrefaction performed in inert atmosphere at 200–300 °C, the same physical and chemical properties of raw biomass change as follows: its moisture decreases, energy density and heating values increase, O/C and H/C ratios decrease, it becomes hydrophobic, its grindability becomes better and its properties change from heterogeneous to more homogeneous. Different names for torrefaction include: biomass carbonization, roasting, slow- and mild pyrolysis, wood cooking and high-temperature drying. Its mechanism based on hemicellulose, lignin and cellulose thermal degradation has been clearly explained in.

Biochar

Similar to charcoal, biochar (also known as biocarbon, biocarbonate or BIOcarbon) is obtained during the slow heating of biomass in absence of air. This process is currently carried out in closed batch heating reactors working either in periodic or continuous mode. The course of biomass biocarbonization reaction with the general formula $(CH_2O)_n$ is as follows:

$$(CH_2O)_n \rightarrow C + H_2O + tar + volatiles + gases$$

Biochar has a much higher calorific value compared to dry biomass from which it was created, due to it no longer containing water physically and chemically related to biomass. For example, biocarbon from pine chips with a heating value of 19.6 MJ/kg has a heating value of 27.9 MJ/kg, from alder shavings (18 MJ/kg)–26.7 MJ/kg, from furniture waste (17 MJ/kg)–26 MJ/kg, from chipboard (17.2 MJ/kg)–25.4 MJ/kg and from fine sawdust (16.3 MJ/kg)–26.6 MJ/kg.

Torrefaction Reactors and Technologies

The reactors for biomass slow (carbonization) and fast biomass pyrolysis are very similar, and sometimes even identical. As all devices for anaerobic, thermal conversion of biomass, these reactors must be hermetic to avoid any uncontrolled supply of air, which may burn the biochar, destroy the installation and even cause a fire. Depending on the heating method, both torrefaction and pyrolysis reactions can be divided into two types: with direct and indirect modes. In the first mode, the energy is transferred by heating media (flue gasses, heat transfer material (HTM) etc.) in direct contact with the feedstock. In the second, indirect mode heating application separates the heating media from the biomass by a conductive barrier such as metal walls. Biomass can also be heated by microwaves at a frequency of 2.4 GHz. This frequency forces molecules of biomass to oscillate at the resonant frequency and induces friction and heating.

From a technological standpoint, torrefaction reactors are divided into reactors operating periodically and continuously. Because periodical reactors only have historical or research (laboratory) significance, one should focus on continuous industrial reactors, such as auger, rotary drum, multistage, multiple hearth furnace (MHF), microwave and others. Torrefaction reactors can also be classified depending on how the batch is forced to move inside them, i.e., mechanically (auger, rotary, ablative), gravitationally (multiple hearth furnace, turbo-dryer, moving compact bed) and pneumatically (torbed, fluidized bed). Descriptions and diagrams of some of these reactors are presented below.

Auger Reactor

The first commercial torrefaction installation operating continuously, employed in the Pechiny process in 1980s (France) included an auger reactor. This reactor consists of one or more helix screws (augers) attached to the shaft rotating in the center of a cylinder. The feedstock moved within the cylinder by the auger mechanism is mixed and, in this case, indirectly heated by cylinder walls. The cylinder containing the feedstock is wrapped in a co-centered, fixed electrical heater with spherical thermal insulation, or a separate pipe (of larger diameter) with hot flue gas medium flowing inside. Gasses generated by the reaction are separated from the solid stream at the end of the screw reactor.

A pilot scale plant (200 kg/h) included a pre-heater rotary drum (t = 220 °C) and an auger screw reactor for continuous process of torrefaction performed in N_2 atmosphere at t = 308 ± 2 °C. Residence time of 9 min yielded 82% efficiency for willow and 80% efficiency for forest residues. The same auger reactor used for torrefaction of biomass, but for a feed rate of 60 kg/h, temperature t = 265 °C, and retention time of 30 min is described in. The auger reactors have screws with mixing paddles and cut flighting, which intensify mixing and heat transfer.

Rotary Drum Reactor

A rotary drum reactor is adapted for torrefaction from drying technology. It is a long, rotating hollow cylinder placed horizontally with a slight angle of inclination for feedstock flow. The rotation of cylinder mixes the material inside and enables intensive heat transfer directly from inert gas or indirectly from the reactor wall. The directly heated reactor utilizes super-heated steam or hot flue gasses from a combustion chamber to create a hot environment of torrefaction. Indirect heating utilizes burners or electric heaters that heat the outer cylinder walls. The reactor with a length of 1.7 m and a diameter of 0.27 m, electrically heated by five external spherical heaters, was used in a pilot plant with a max. production output of 20 kg/h.

Multiple Hearth Furnace (MHF)

The MHF is constructed from round, multi-leveled platforms with internal heating mechanism, steam injectors or gas burners. The torrefacted biomass is supplied from the top of the furnace and gravitationally falls from the round stationary shelf to the rotating shelfs through holes in the center of the stationary shelf. In the first variant (electrically heated reactor with conical shelfs), the angle at the base of the shelves is equal to or greater than the bulk angle of the crushed biomass. This enables automatic gravitational biomass movement from shelf to shelf. In the second version (reactor heated by exhaust gases), biomass movement is caused by scraper blades alternately attached to the rotating shaft and the cylindrical wall of the reactor. Rotary blades scrape the biomass from the stationary shelves towards the center, where the load falls onto the rotary shelf through the central holes. Subsequently, fixed shovels scrape the biomass from the rotary shelves onto the stationary shelf.

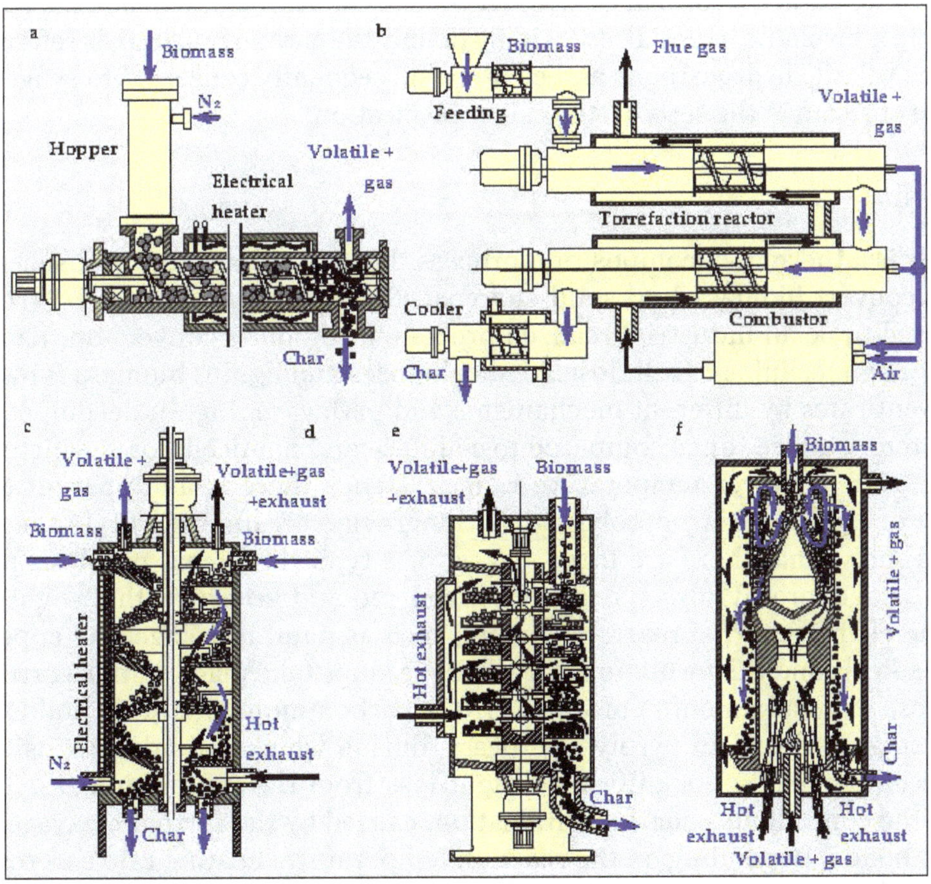

Diagrams of the most popular reactors and devices for continuous biomass torrefaction.

A reactor with gravitational movement of the charge (Standard TURBO-DRYER). In this column, toroid round shelves are attached to a rotary hollow shaft. Flue and gas torrefaction products are sucked in through the holes above the shelves and are transported upwards through the channel in the middle of the shaft to the outside of the reactor. The biomass supplied from above is directed by scrapers attached to the walls of the reactor into radial holes in the shelf, through which the charge falls down onto a lower shelf. This operating principle can be duplicated as many times as needed to achieve a certain processing level. Independent heating of each shelfs enables better continuous process temperature control so that after drying taking place in the upper part of the reactor, torrefaction can occur in the lower part.

Pyrolysis and Gasification

The chemistry of biomass gasification is very complex. Biomass gasification proceeds via a two-step process; pyrolysis followed by gasification that includes gas and tars reforming plus char and soot conversion. Pyrolysis is the decomposition of the biomass by heat. This step, also known as devolatilization, is endothermic and produces mainly volatile materials in the form of gaseous and condensable hydrocarbons called tars. The remaining nonvolatile material, containing mainly the carbon material, is referred to as char. The volatile hydrocarbons and char are subsequently converted to syngas in the presence of steam in the second step called gasification.

Pyrolysis

Pyrolysis is a thermal decomposition process that takes place in the absence of oxygen to convert biomass into solid charcoal, liquid (bio-oil) and gas. Pyrolysis is considered to be an industrial realized process for biomass conversion. Each component of lignocellulosic (cellulose, hemicelluloses and lignin) biomass is pyrolysed at different rates by different mechanisms and pathways. Lignin decomposes over a wider temperature range compared to cellulose and hemicelluloses which rapidly degrade over narrower temperature ranges. Hence there is an apparent thermal stability of lignin during pyrolysis. Thermogravimetry analysis (TGA) testing of biomass shows that there are three stages for a typical biomass pyrolysis process. The first stage, pre-pyrolysis, occurs between 120 and 200 °C with a slight weight loss, when some internal rearrangements, such as bond breakage, the appearance of free radicals and the formation of carbonyl groups take place, with a corresponding release of small amounts of water (H_2O), carbon monoxide (CO) and CO_2. The second-stage is the main pyrolysis process, during which solid decomposition occurs, accompanied by a significant weight loss from the initial biomass. The last stage is the continuous char devolatilization, caused by the further cleavage of C-H and C-O bonds. Depending on the reaction temperature, heating rate and residence time, pyrolysis can be classified into slow pyrolysis (intermediate pyrolysis) fast pyrolysis and flash pyrolysis.

Slow Pyrolysis

Slow pyrolysis has been applied for thousands of years and has been mainly used for the production of charcoal. In slow pyrolysis, biomass was typically heated about 500 °C at slow heating rates (up to 10 - 20 °C/min). The vapour residence time varies from 5 min to 30 min. Thus, the components in the vapour phase continue to react with each other, as solid char and liquid are being formed. The main product, charcoal, can be used in a wide range of areas, from domestic cooking and heating to metallurgical or chemical use as the raw material for production of chemicals, activated carbon, fireworks, absorbents, soil conditioners and pharmaceuticals. A higher yield of charcoal can be obtained from biomass feedstocks with higher lignin contents and lower hemicelluloses contents. In contrast to fast pyrolysis, slow pyrolysis does not necessarily require fine feedstock particle size (smaller than 1 mm).

Fast Pyrolysis

Fast pyrolysis is a process in which very high heat flux are imposed to biomass particles, leading to very high heating rates, in the absence of oxygen. Biomass decomposes to generate vapours, aerosol and char. After cooling and condensation of the vapours and aerosol, a dark brown mobile liquid is formed which has a heating value of about half of the conventional fuel oil. Fast pyrolysis process produces 60-75 wt% of liquid bio-oil, 15-25-wt% of solid char and 10-20-wt% of non-condensable gas, depending on the feedstock used. No waste is generated because the bio-oil and solid char can each be used as a fuel and the gas can be recycled back in the process. Fast pyrolysis uses much higher heating rates than slow pyrolysis. While slow pyrolysis is related to the traditional pyrolysis processes for making charcoal, fast pyrolysis is an advanced process which is carefully controlled to give high yields of liquid. Research has shown that maximum liquid yields are obtained with high heating rates, at reaction temperatures around 500 °C and with short vapour residence times to minimize secondary reactions. Very short residence times result in incomplete depolymerization of the lignin due to random bond cleavage and inter-reaction of the lignin macromolecule resulting in a less homogenous liquid product, while longer residence times can cause secondary cracking of the primary products, reducing yield and adversely affecting bio-oil properties. The essential features of fast pyrolysis process are:

- Very high heating and heat transfer rates, which usually requires a finely ground biomass feed: <1mm.
- Carefully controlled pyrolysis reaction temperature of around 500 °C in the vapour phase, with short vapour residence times of typically less than 2 s.
- Rapid cooling of the pyrolysis vapours to give the bio-oil product.

Flash Pyrolysis

Very fast pyrolysis is sometimes referred to as "flash pyrolysis", usually in the context of laboratory studies involving rapid movement of substrate through a heated tube under gravity or in a gas flow. Higher temperatures and shorter residence times than fast pyrolysis are used; the main product distributions are similar to fast pyrolysis. The distinction between flash and fast pyrolysis has largely disappeared and now the term "flash" has largely disappeared and is gradually being replaced by a more generalized definition for fast pyrolysis.

Fast Pyrolysis Reactor Configuration

Fast pyrolysis conversion technology has led to design of original reactor systems that provide the essential ingredients of high heating rates, moderate temperatures and short vapour product residence times for liquids. The most commonly used reactors for fast pyrolysis are bubbling fluidized-bed, circulating fluidized-bed, ablative, entrained flow, rotating cone and vacuum reactors. There are three main methods achieving fast pyrolysis.

- Ablative pyrolysis, in which heat is transferred by conduction. Wood is pressed against a heated surface and rapidly moved during which the wood melts at the heated surface and leaves an oil film behind which evaporates. This process uses large particles of wood and is typically limited by the rate of heat supply to the reactor. It leads to compact and intensive reactors that do not need a carrier gas, but with the penalties of surface area controlled system and moving parts at high temperature.
- Bubbling fluidized bed and circulating fluidized bed pyrolysis, in which heat is transferred from a heat source to the biomass by a mixture of convection and radiation. The heat transfer limitation is within the particle, thus, requiring very small particles of typically no more than 3 mm to obtain good liquid yields. Substantial carrier gas is needed for fluidisation or transport.
- Vacuum pyrolysis, which has slow heating rates but removes pyrolysis products as rapidly as in the previous methods, which thus simulates fast pyrolysis. Larger particles can be accepted but the vacuum leads to larger equipment and higher costs. Total liquid yields are typically lower than 60-65 wt% compared to 75-80 wt% with the previous two methods.

Pyrolysis Products

- Char: Char is a porous carbon structure that remains after the hydrogen and oxygen fractions have left the fuel. Char is often defined as the solid residue after pyrolysis. It is often polluted with other components: mineral fractions and after incomplete pyrolysis, large fractions of hydrogen and oxygen, that can still be present in char. Char is believed to contribute to the formation of polycyclic aromatic hydrocarbon (PAHs) during biomass pyrolysis, particularly at low temperature. Char can be used as a fuel in form of briquettes or as a char-oil, charwater slurry; alternatively char can be upgraded to activated carbon and used in purification processes. The properties of the char obtained after biomass pyrolysis have a direct influence on the subsequent char oxidation step, since the amount and type of pores determine the gas accessibility to the active surface sites. Properties of char are decisively affected not only by properties of parent material but also by pyrolysis operating conditions, mainly the heating rate, the maximum temperature and the residence time at this temperature.

- Pyrolysis liquid: The liquid product from biomass pyrolysis is known as bio-oil. Bio-oil is not a product of thermodynamic equilibrium during pyrolysis but, it is produced with short residence times and rapid cooling or quenching from the pyrolysis temperature. This condensate is not at thermodynamic equilibrium at storage temperatures. Hence the bio-oil chemical composition tends to change toward thermodynamic equilibrium during storage.
- Gas: The third main product from pyrolysis is gas. The gas mainly consists of H_2, CO_2, CO and CH_4 together with traces of C_2 species. CO_2 and CO are preferentially produced at low temperature, while H_2 is produced at high temperature.

Pyrolysis Reaction Schemes

The exact pyrolysis mechanisms of biomass are still unknown, although substantial literature sources are available on biomass devolatilization kinetics and mechanisms. Many biomass devolatilization models have been developed. One-step global mechanisms and semi-global multi-step mechanisms can be basically distinguished. The simplified approaches define devolatilization rates with single or two-step Arrhenius reaction schemes involving pseudo-species. The one-step global mechanisms can be shown as:

$$\text{Biomass} \xrightarrow{k} \text{Volatiles} + \text{Char} \quad \text{or}$$

$$\text{Biomass} \xrightarrow{k} \text{Tar}(\text{Bio}-\text{oil}) + \text{Gas} + \text{Char}$$

The reaction kinetic rate (k) is expressed in single-step Arrhenius law form as k =A exp(-Ea/RT) and the devolatilization rate is expressed as:

$$\frac{-dm_p}{dt} = k\left[m_p - \left(1 - f_{v,0}\right)m_{p,0}\right]$$

where m_p is the biomass particle mass, $m_{p,o}$ is the initial particle mass and $f_{v,o}$ is the initial volatile fraction. For two-step Arrhenius reaction schemes, the kinetic devolatilization rate expressions of the form are:

$$k_1 = A_1 \exp(-Ea/RT)$$

$$k_2 = A_2 \exp(-Ea/RT)$$

where k_1 and k_2 are competing rates that may control the devolatilization over different temperature ranges. One-step multi-reaction schemes have been developed via three parallel reactions into char, tar and gas and can be shown as follow:

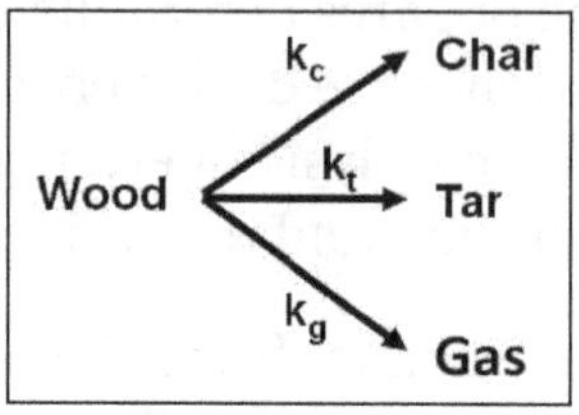

More recent models are adapted to be able to handle different feedstocks. The composition of the feedstock is represented using three model species. In these models the source species like wood or biomass is replaced by a mixture of cellulose, hemicelluloses and lignin. Most models are only proposed for cellulose, but several schemes were also proposed for the other biomass main components. Several other kinetic schemes for cellulose pyrolysis are proposed below. Researchers have developed a multistep model. In this scheme, cellulose is firstly converted to an active state by an initiation reaction. Hereafter the active cellulose reacts via two competing reactions to produce tar, char and gas.

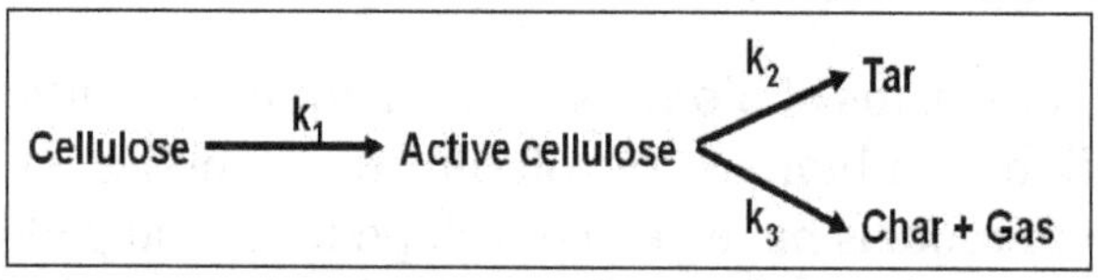

The two competing reactions represent two pathways that the decomposition of cellulose can follow. The "char path" is favored at low temperature (200-280 °C) whereas the "tar path" dominates at elevated temperatures (280-340 °C).

Secondary Reactions

Tars produced during the decomposition of the virgin fuel can decompose further. Several authors suggest a transformation of tar to char and gas following two independent reactions. Tar decomposition is suggested to be catalyzed by the solid matrix of the fuel resulting in char creation on the pore walls.

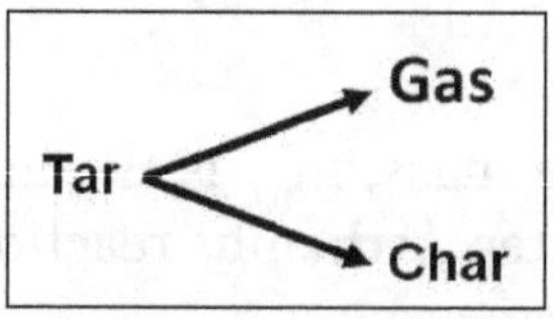

It is unclear however how this takes place. A single tar reaction with only gas as product species.

$$\text{Tar} \longrightarrow \text{Gas}$$

Gasification

Gasification Chemistry

Gasification can be seen as an extension of pyrolysis; biomass gasification is a complex thermochemical process involving numerous different reactions. The biomass gasification process can be divided into two parts: pyrolysis and gasification. In the gasification part the gas, tar and char react further. Gasification, as a core technology for the production of chemicals and clean power, refers to a process converting biomass into either fuel gas (containing CH_4 and some N_2 usually) or syngas (containing mainly H_2 and CO). During gasification several reactions take place. A simplified representation of these reactions is given in equations.

Heterogeneous reactions:

$$C_{(s)} + O_2 \leftrightarrow CO_2$$

$$\ddot{A}H^{o}_{298} = \text{-}394\ \text{kJmol}^{-1}$$

$$C_{(s)} + H_2O \leftrightarrow CO + H_2$$

$$\ddot{A}H^{o}_{298} = 131\ \text{kJmol}^{-1}$$

$$C_{(s)} + CO_2 \leftrightarrow 2CO$$

$$\ddot{A}H^{o}_{298} = 173\ \text{kJmol}^{-1}$$

$$C_{(s)} + 2H_2 \leftrightarrow CH_4$$

$$\ddot{A}H^{o}_{298} = \text{-}75\ \text{kJmol}^{-1}$$

Homogeneous reactions:

$$CO + H_2O \leftrightarrow CO_2 + H_2$$

$$\Delta H^{o}_{298} = -41\ \text{kJmol}^{-1}$$

$$CH_4 + H_2O \leftrightarrow CO + 3H_2$$

$$\Delta H^{o}_{298} = 206\ \text{kJmol}^{-1}$$

Cracking reaction:

$$\text{Tar} \rightarrow aCO + bCH_4 + cC$$

Three major reaction regimes during the gasification process identified as primary, secondary and tertiary regimes as shown in figure. Biomass is mainly converted into a mixture of hydrogen, carbon monoxide, light hydrocarbons, such as methane and other non-combustible gas such as carbon dioxide, water vapour and nitrogen. In most cases, the product gas can contain particulates such as char, ash, soot, etc. Reactor temperature is one of the most important operating conditions which affect both the heating value and producer gas composition. The gasification process is usually performed with aid of a gasification agent. The gasification agent can be steam, or air, or enriched air, or oxygen, or a combination of steam and an oxygen source, or carbon dioxide. The process is performed at relative high temperature 600- 1500 °C. There are also small amounts of impurities in the gas: char, soot, tars, alkalis, nitrogen compounds, sulphur compounds and chlorine compounds. The composition of the product gas can vary significantly depending on operating conditions (e.g. temperature, operating pressure and oxidant agent), type of feed stock, moisture content in the fuel, mode of bringing the reactants into contact inside the gasifier. Moreover, the quality of the product gas is associated with other factors such as the type of gasifier, the residence time and the heating rate which is usually associated with the type of feed stock, particle size and temperature.

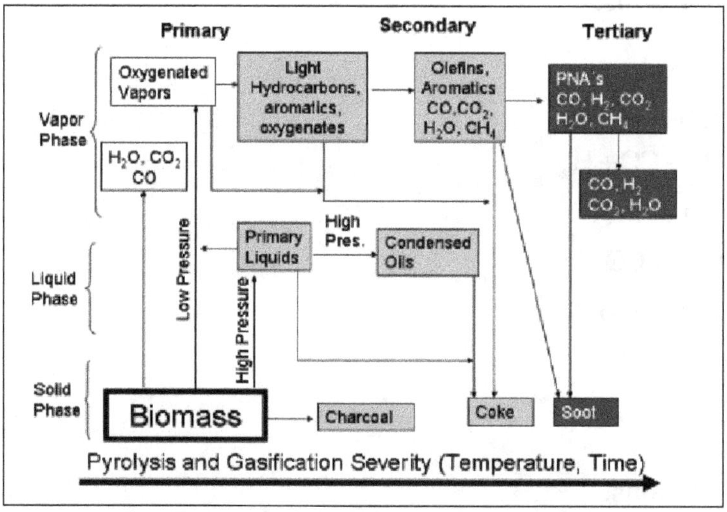

Gasification and pyrolysis reaction pathways.

There are hundreds of different types of gasifiers in the patent literature. However, they can be divided into four main types: Updraft gasifier, Downdraft gasifier, Fluidized-bed gasifier and Entrained flow gasifier.

Types of Gasifier

Until now, most gasifiers have been developed and commercialized for the production of heat and power from syngas. The main differences between gasifiers are:

- The type of injection: Biomass is either fed from the top of the gasifier or from the side and then is moved around either by gravity or air flows.
- The gasification agent used.
- The type of heating: It can be done either by partial combustion of the biomass in the gasifier (directly heated), or from an external source (indirectly heated), such as circulation of an inert material.

- The temperature range.
- The pressure range under which the gasifier is operated.

Updraft Gasifier

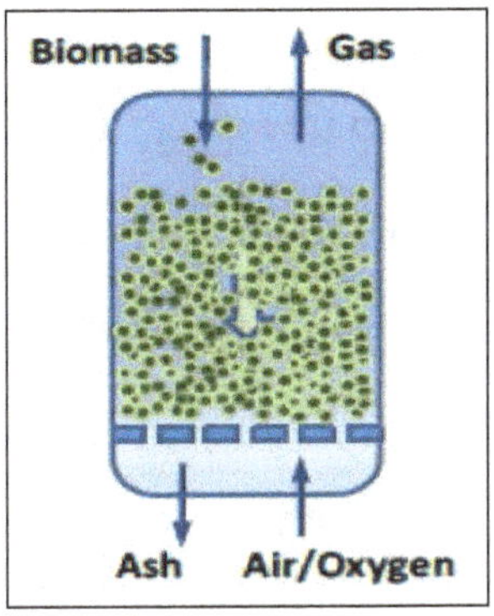

The Updraft gasifier (counter-current reactor) scheme.

Updraft reactor is the simplest and oldest form of gasifier, also known as counter-current gasifier. Biomass enters at the top of the reactor and air/oxygen/steam enters at the bottom of the reactor, flows upward and the product gas leaves at the top. In figure, a schematic representation of an updraft gasifier is given. Complete combustion of char takes place at the bottom of the bed, liberating CO_2 and H_2O. This hot gas (~1000 °C) passes through the bed above, where it is reduced to H_2 and CO and cooled to 750 °C. Going up the reactor, the reducing gas (H_2 and CO) pyrolyses the descending dry biomass and finally dries the incoming wet biomass, leaving the reactor at a quite low temperature (~500 °C).

Downdraft Gasifier

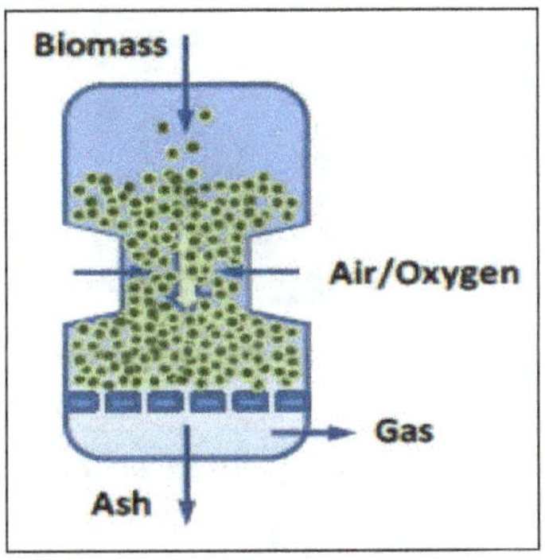

The downdraft gasifier (co-current reactor) scheme.

This gasifier was widespread during World War II. It was used to convert biomass or charcoal into fuel for gasoline vehicles in periods of great shortages of transportation fuels. The downdraft gasifier is a co-current reactor, where the fuel and the gasification agent move in the same direction. In figure, a schematic representation of a downdraft gasifier is given. There are different reaction zones in a downdraft gasifier. In the drying zone, moisture is evaporated from biomass as it slowly moves down towards the pyrolysis zone. In the pyrolysis zone, biomass is converted into char, tars and gas. Some of the pyrolysis products are combusted. Due to the high temperature, tars are cracked. As a result, the produced gas is relatively clean. However, downdraft gasifier has some drawbacks:

- It requires feedstocks drying to low moisture content.
- Syngas exiting the reactor is at high temperature, requiring secondary heat recovery system.

Fluidized-bed Gasifier

The biomass, which is previously reduced to a fine particle size and air, steam, or oxygen enter at the bottom of the reactor. An inert or catalytic bed material is used to transport heat and mass through the reactor. The bed is fluidized by blowing the gasification agent through the bed, which lifts the bed against gravity. As a result, the turbulence in the bed creates an even temperature distribution in the bed. In figure, a schematic representation is given of a fluidized-bed gasifier. Therefore, contrary to fixed bed gasifiers there are no different reaction zones in fluidized bed gasifiers. The operating temperature of a fluidized bed is usually of 700-900 °C and the pressure range is between 0 and 70 bar. The most common types of fluidized beds are the bubbling fluidized bed (BFB) and the circulating fluidized bed (CFB).

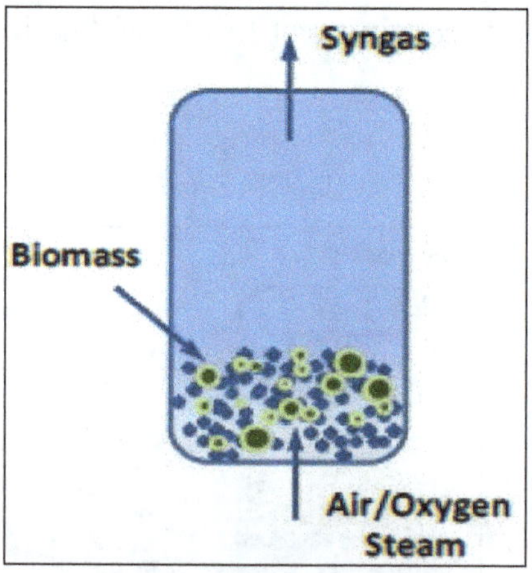

Fluidized-bed gasifiers scheme.

The advantages of fluidized-bed gasifiers are:

- Exhibit a nearly uniform temperature distribution throughout the reactor.
- Provide high rates of heat transfer between inert material, biomass and gas.
- High conversion possible with low tar and low unconverted carbon.

The disadvantages of fluidized-bed gasifiers are:

- The tolerance of high moisture content feedstocks is much more limited.
- The system design is also more complex, requiring blowers to inject the oxidant at the bottom of the fluidized bed.
- Bed sintering when a biomass with high ash content is used. The alkali components in the ash have the tendency to form low-melting eutectics with silica being the most common bed material, which can lead to loss of fluidization.

Entrained Flow Gasifier

In an entrained flow gasifier, as shown in figure, the biomass is present in dust form. Contact surfaces are very high and hence reaction times are very short. The feed and air move co-currently and the reactions occur in a dense cloud of particles at very high temperature >1000 °C so that firstly the syngas results almost free from tars and secondly ash melts, being then collected at the bottom of the reactor in the form of slag. Conversion in entrained flow reactors is close to 100%. There is little experience with biomass in such systems.

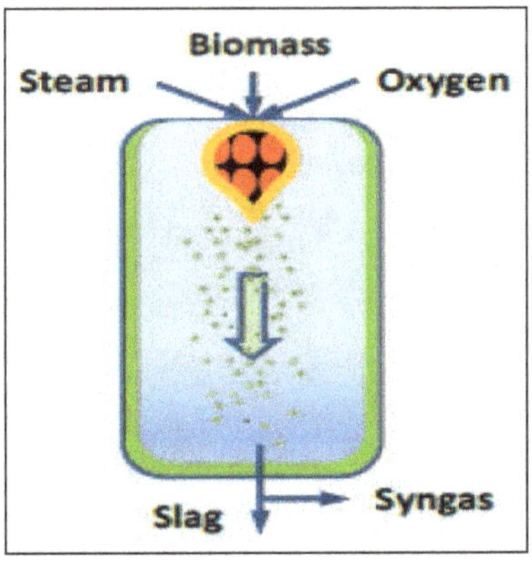

Entrained flow gasifier schematic.

Syngas

Syngas is always confused with the producer gas, however there is a difference between those two terms:

- Producer gas is the mixture of gas produced by the gasification of biomass at relatively low temperatures (700 to 1000 °C). Producer gas is composed of carbon monoxide (CO), hydrogen (H_2), carbon dioxide (CO_2) and typically a range of hydrocarbons such as methane (CH_4). Producer gas can be burned as a fuel gas such as in a boiler for heat or in an internal combustion gas engine for electricity generation or combined heat and power. The composition of the gas can be modified by change in of gasification parameters.

- Syngas (synthesis gas) is a mixture of carbon monoxide (CO) and hydrogen (H_2), which is the product of high temperature steam gasification of biomass. Following clean-up to remove any impurities such as tars, syngas can be used to produce organic molecules such as synthetic natural gas or liquid biofuels such as synthetic diesel or other products.

There are four main uses of syngas that are currently being explored for production of liquid fuels:

- Fischer-Tropsch synthesis, a chemical catalytic process that has been used since the 1920s to produce liquid fuels from coal-derived syngas and natural gas.
- Mixed alcohols synthesis, a chemical catalytic process that produces a mixture of methanol, ethanol, propanol, butanol and smaller amounts of heavier alcohols.
- Syngas fermentation, a biological process that uses anaerobic microorganisms to ferment the syngas to produce ethanol or other chemicals.
- Methanol synthesis, also a chemical catalytic process currently used to produce methanol. Figure shows routes for transportation fuels and chemicals production from syngas.

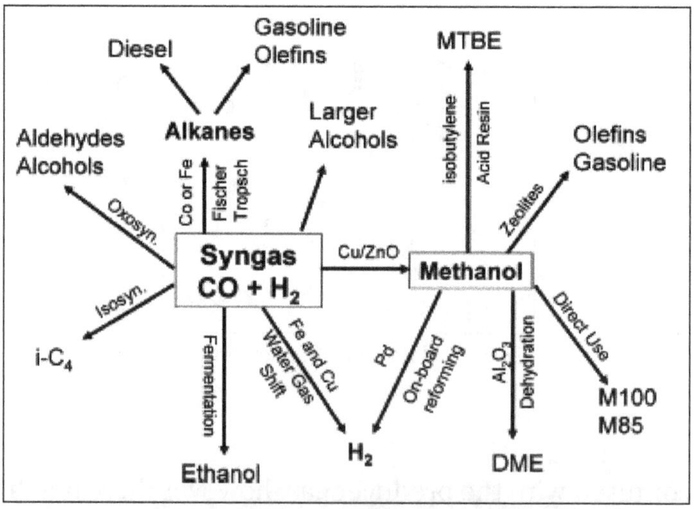

Pathways for fuel production from syngas.

For energy production, the major concerns about syngas are its heating value, composition and possible contamination. The composition of syngas depends on the biomass properties and gasifier operating conditions. The principle contaminants in syngas, as it leaves the gasifier, are: soot, ash, tars, H_2S (hydrogen sulphide) depending on the sulphur content of the feed and trace quantities of NH_3 (ammonia), COS (carbonyl sulphide), HCl (hydogen chloride), HCN (hydrogen cyanide), and heavy metals.

Biochemical Conversion

Biogas Formation

Biochemical conversion of biomass includes utilization of bacteria, microorganisms and enzymes to break biomass into gaseous or liquid fuels, such as biogas or bio-ethanol. The most popular biochemical conversion technologies are:

- Anaerobic digestion (or bio-methanation).
- Ethanol fermentation.

Anaerobic Digestion

Anaerobic digestion is extensively utilized for recycling and treating wet organic waste. It is a commercially proven technology. It comes in category of fermentation that transforms organic substance into biogas, which mostly consists of methane (approximately 60%) and carbon dioxide (approximately 40%) and is equivalent to landfill gas.

Anaerobic digestion is a biologically natural method of conversion that occurs when bacteria breaks into biodegradable material in the absence of oxygen, which creates biogas, a mixture of mostly methane and carbon dioxide with small quantities of other gases like hydrogen sulphide.

By this process biomass is converted into cattle dung, human wastes and other organic waste with very high moisture content into biogas (gobar gas) through anaerobic fermentation in absence of air. Fermentation occurs in two levels by two different metabolic groups of bacteria. Initially the organic material is hydrolyzed into fatty acids, alcohols, sugars, H_2 and CO_2. Methane forming bacteria then converts the product of first stage to CH_4 and CO_2 in the temperature range 30-55 °C. Fermentation occurs in a sealed tank called 'digester' where the sludge is left behind is used as enriched fertilizer.

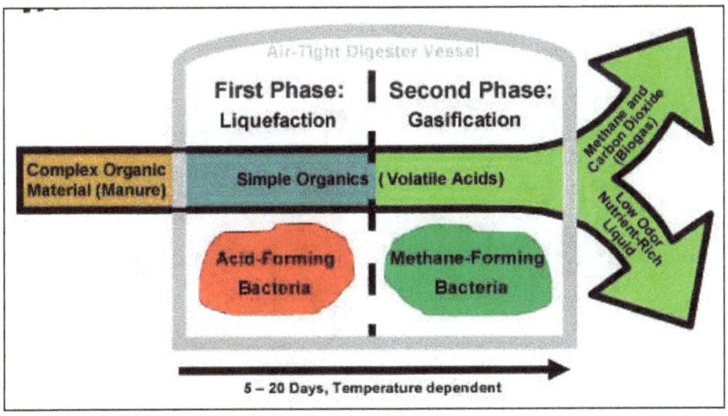

Working of Anaerobic Digestion Process.

Table: Characteristics of Anaerobic Digestion.

Characteristic	Anaerobic
Reaction	$C_6H_{12}O_6 \rightarrow CO_2 + 3\,CH_4$
Energy Release	$\Delta G^{o''}$ = -393 kJ/mol glucose
Carbon Balance	95% $\rightarrow CO_2 + CH_4$ (= biogas)5% → biomass
Energy Balance	90% retained in $CH_4$5% → biomass 5% → heat production
Biomass Production	Slow growth of biomass

Anaerobic digestion plant produces two products, biogas and digestate both can be further processed or used to produce secondary products. Biogas can be used for producing electricity and heat, as a natural gas substitute and also a transportation fuel. Biogas can also be upgraded and used in gas supply networks.

Digestate can be further processed to produce liquor and a fibrous substance. The fibre, which can be processed into compost, is a bulky material with low levels of nutrients and can be used as a soil conditioner or a low level fertilizer. A high proportion of the nutrients remain in the liquor, which can be used as a liquid fertilizer.

Power Generation from Biogas: Bio Power

There are three methods for this:

- Direct method
- Gasification
- Landfill

Direct Method

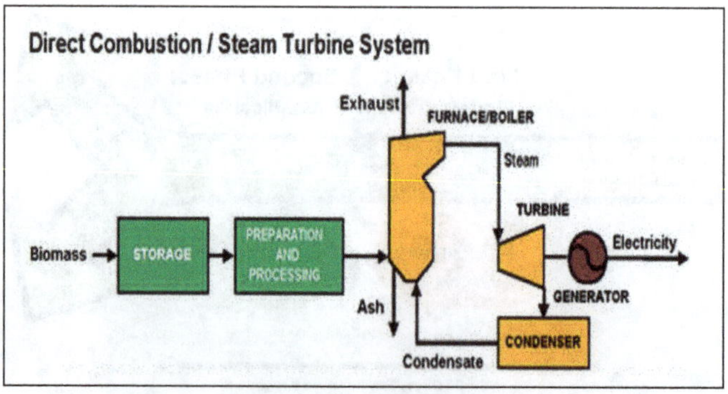

Direct method.

For decades, the only commercial options available for the production of electricity are wood, wood residues and other solid. In the method biomass is directly combusted and coupled to a steam turbine – that is, burning biomass in a boiler to produce hot gases, producing steam from the hot gases via a heat exchanger, and then generating power from the steam using a steam turbine.

Combustion of the fuel produces steam in the boiler that powers a turbine/generator to generate electricity. The steam is then returned to the liquid state in a condenser by circulating cooling water around the condenser tubes containing the steam. The condensed water is then recirculated to the boiler to be heated again into steam.

- Combined Heat and Power: Combined heat and power or CHP, also called cogeneration or distributed generation, is the simultaneous production of two types of energy – heat and electricity – from one fuel source, often natural gas.
- Cogeneration with Biogas: Biogas from Regenerative Raw Material and Biogas Plants: Biogas is the product of the anaerobic fermentation of organic material in a fermenter. This gas has a CH_4 methane content of 45-70 percent. During the combustion, the amount of emitted CO_2 merely corresponds to the amount the substrates absorbed during growth. Cogeneration plants that produce power using biogas prevent emissions that would otherwise be emitted during the combustion of fossil fuels. A typical CHP system has an electric generator, which can take the form of a gas turbine, steam turbine, or combustion engine. In addition to this electric generator, a waste heat exchanger is installed, which recovers the excess heat or exhaust gas from the electric generator to in turn generate steam or hot water.

Gasification Technology

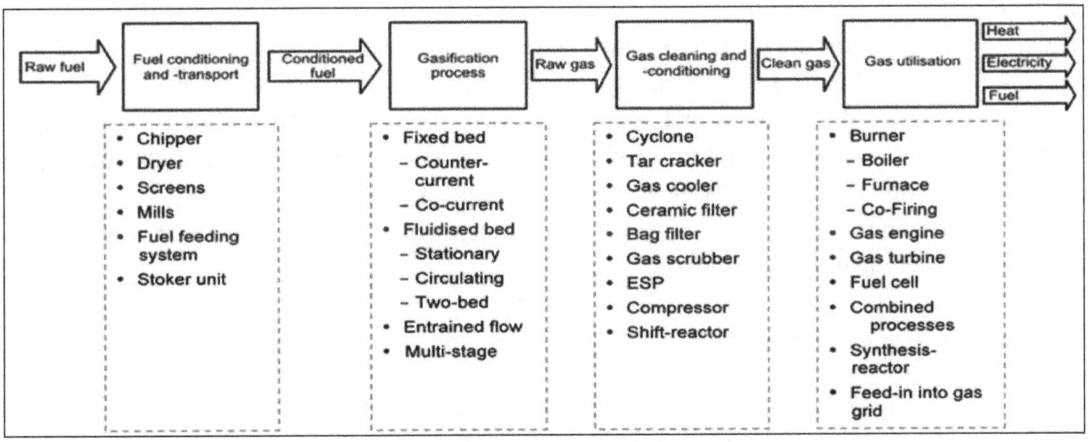

Processes involved in gasification.

Gasification plants consist of several process steps. The solid biomass fuel delivered needs to be adjusted (fuel conditioning and handling) to the fuel characteristics (particle size, water content) required for the gasification process. The conditioned fuel enters the gasification process, which produces raw product gas. The raw product gas quality needed for further utilization. The cleaned product gas is used for the production of electric power, heat and fuel based on different technologies.

Landfill

Generally during anaerobic conditions, the composition of landfill gas is approximately 50 percent methane and 50 percent carbon dioxide with trace amounts of nitrogen, oxygen, hydrogen sulphide, hydrogen, and non-methane organic compounds (NMOCs). The more organic waste and moisture present in a landfill, the more landfill gas is produced by the bacteria during decomposition. The process involves covering or "capping" a section of the landfill with clay, drilling vertical wells into the waste and connecting them with underground piping. Vacuum compressors then capture the gas and pipe it to a blower/flare station, where it is either piped to the power generation facility or combusted in a flare. During the combustion process the methane is converted to carbon dioxide, a less harmful gas. While effective at reducing methane gas emissions, ending the process at this point makes no use of the heat energy that is produced during the conversion. After the gas is collected, it will be piped to a power generation facility and combusted in engine generator sets to generate electricity.

Ethanol Formation

Pretreatment

The purpose of the pretreatment step is to further increase the surface area of the lignocellulosic material, disrupt the structure of the lignocellulose such that the cellulose component is accessible to hydrolyzing agents and reduce the crystallinity of the cellulose to further facilitate hydrolysis. Depending on the nature of the pretreatment technology selected, this step can also include solubilization of the lignin or the hemicellulose component. Various pretreatment options are available now to fractionate, solubilize, hydrolyze and separate cellulose, hemicellulose, and lignin components. These include physical, physicochemical, chemical and biological pretreatment. Consequently, the pretreatment process represents a significant cost element of the whole lignocellulosic bioethanol process.

Feedstock Size Reduction

Before pretreatment, the first stage in the production of ethanol from biomass is cleaning followed by mechanical comminution combines chipping, grinding, and milling to break the lignocellulosic materials down to 0.2 to 2 mm and reduce the crystallinity of the materials. Size reduction is necessary to provide pumpable slurry and to increase the biomass surface area so that mass transfer effects are minimized during the downstream processes. Techniques for size reduction include hammer, disk and knife milling and are well established.

Hydrolysis

In the hydrolysis reaction, the complex chains of sugars that make up the hemi-cellulose are broken, releasing simple sugars. The complex hemi-cellulose sugars are converted to a mix of soluble five-carbon sugars, xylose and arabinose, and soluble six-carbon sugars, mannose and galactose. The rest of hemicelluloses are degraded to weak acids, furan derivates, and phenolics. These compounds, however, are potential fermentation inhibitors. By the action of dilute acids, concentrated acids, and/or enzymes (Cellulase), the glucose yields of cellulose hydrolysis often exceed 90%, but hydrolysis without preceding pretreatment yields typically less than 20% only. The cellulose hydrolysis reactions can be simply represented as:

$$\left(C_5H_8O_4\right)_n + n\,H_2O \rightarrow n\left(C_5H_{10}O_5\right)$$
$$\left(C_6H_{10}O_5\right)_n + n\,H_2O \rightarrow n\left(C_6H_{12}O_6\right)$$

The pretreated feedstock can be hydrolysed by two methods (Acid hydrolysis and Enzyme hydrolysis).

Acid Hydrolysis

In the past various acid hydrolysis technologies have been developed. The different acid hydrolysis technologies can be divided into two broad categories: i) hydrolysis with concentrated acid at low temperatures; ii) hydrolysis with dilute acid at high temperatures. Acid hydrolysis technologies have a long industrial history but lead to high operating costs and various environmental and corrosion problems.

- Dilute acid hydrolysis: The dilute acid process is conducted under high temperature and pressure and has a reaction time at a scale of up to minutes, facilitating continuous processing.
- Concentrated acid hydrolysis: The concentrated acid process uses relatively mild conditions, with a much longer reaction time.

Enzyme Hydrolysis

Another basic method of hydrolysis is enzymatic hydrolysis. Enzymes are naturally occurring plant proteins that cause certain chemical reactions to occur. Enzymatic hydrolysis is not commercialized yet but is recognized to be the most promising hydrolysis technology. A reduction of the cost of ethanol production can be achieved by reducing the cost of either the raw materials or the cellulase enzymes. Reducing the cost of cellulase enzyme production is a key issue in the enzymatic hydrolysis of lignocellulosic materials. Enzymatic hydrolysis of cellulose is usually carried out by cellulase enzymes. During hydrolysis, cellulose is degraded into the reducing sugars that can be fermented by yeasts or bacteria to ethanol.

Fermentation Process

The fermentation of, hydrolyzed product, glucose into ethanol can be carried out using a biocatalyst, called Saccharomyces cerevisiae yeast or Zymomonas mobilis bacteria. Saccharomyces cerevisiae and related species have the ability to utilize a wide range of hexoses such as glucose, fructose, sucrose, galactose, maltose and maltotriose to produce a high yield of ethanol. The fermenting of the biomass is conducted under standard fermenting conditions and will utilize all the major biomass. Yeasts are minute, often unicellular, fungi. The yeasts used are typically brewers' yeasts. Examples of yeast capable of fermenting the decaying biomass include, but are not limited to, Saccharomyces cerevisiae and Saccharomyces uvarum. Non-Sacharomyces yeasts, also known as non-conventional yeasts, are also used to make a number of commercial products. Some examples of non-conventional yeasts include Kuyberomyces lactis, Yarrowia lipolytica, Hansenula polymorpha and Pichia pastoris. Microorganisms other than yeast can also be useful in making fermentation products. For example, cellulosic ethanol production also utilizes fungi and bacteria. Examples of these cellulolytic fungi include Aspergillus niger, Trichoderma reesei , Trichoderma longibrachiatum and Trichoderma viride. One example of a bacteria used in cellulosic ethanol production is Clostridium thermocellum , Clostridium cellulovorans and Clostridium Ijungdahlii. Mid- to long-term technology under development are expected to improve the fermentation efficiency of the organism, producing higher yields in less time, and an organism requiring less detoxification of the hydrolysate. This process has the advantage of being able to maintain a much higher cell density in the fermentor, thereby increasing ethanol productivity.

Distillation

Separation of ethanol from the fermentation solution refers to the stage in which once ethanol begins to form during fermentation, it is isolated from the fermentation solution. The fermentation solution is likely to contain water, ethanol, and the remaining biomass. Distillation was one of the earliest separation techniques used by alchemists and pharmacists. And, generally, distillation, along with chromatography and filtration, is still considered to be a key method of separating and purifying substances. The separated ethanol, which will generally not be fuel-grade, can be concentrated to fuel grade (at least 95% ethanol by volume) via a second distillation.

A process for producing and recovering light alcohols, particularly ethanol, alcohol mixtures containing ethanol, and ABE mixtures (alcohol mixtures containing acetone, ethanol and butanol), using a combination of steps including fermentation, first membrane separation, dephlegmation and dehydration by second membrane separation.

Ethanol Production Cost

The cost of ethanol production and its value depends on plant location, feedstock, production scale, and end use. It also depends on the availability of feedstock, plant location, feedstock transportation cost, method of pretreatment, hydrolysis, fermentation techniques and ethanol market price. Currently, the most important feedstock for the production of ethanol is sugarcane juice in Brazil and corn in the USA. Other starch-rich grains, such as wheat and barley, are mostly used in Europe. Most of the ethanol produced from North America is from cereal grains such as corn and wheat. Improvements in pretreatment processes and breakthrough in enzyme technology will have an impact on the competitive industrial production of fuel ethanol in processes such as continuous process without or with cell recycling. Simultaneous saccharification and co-fermentation, utilization of immobilized cells, etc. enable higher yield of ethanol and reactor productivity. Economic conversion of cellulosic biomass to ethanol will also reduce the production cost that can be achieved by making ethanol fermentation process faster, particularly xylose fermentation, reducing the formation of by-products and developing genetically engineered yeasts to produce various high-value by-products.

Biomass Heating Systems

Stem Wood and Logs

- Harvesting and Forwarding: Harvesting is performed by using a harvester (a special machine, which grabs the tree and cuts it close to the roots) or manually by using a chain saw. Harvesters are preferred for large scale production as they greatly increase production rates, provide a much higher level of personnel safety, and minimize the amount of persons needed for the harvesting operation. Modern harvesters are very advanced machines, capable of accessing difficult areas while maintaining good operator safety and comfort. They are also equipped with advanced technology for measuring the size of the tree and determining the most suitable log lengths according to the end users' requirements. The stem of the tree will be used by sawmills, pulp industries or heat and power plants and is separated from the remaining parts (residues). This is performed by cutting away the top of the tree and all branches from the stem. The stem is then cut into log lengths that allow easy transport by road and which suit the processing equipment available at the end users site.
- A forwarder is used to transfer logs from the harvesting site to a nearby road where they are stacked to await collection by log trucks. Forwarders are based on the same technology as harvesters, but are more simple and equipped with a

crane and trailer for collection and carrying the logs or residues. They can also be used to transfer residues to the road side where they are stacked to enable drying. The logs can be stored in the forest (or at a terminal) for drying reasons.

A typical stack of logs close to the clear-cut.

Logging Residues

- Harvesting, Chipping, Storage and Bundling: The American Forest Agency has issued recommendations on harvesting methods that should be used in order to help maintain biodiversity in the forest during and after harvesting of logging residues. This has also been discussed in the research programs on forest fuels and their harvesting. According to the recommendations tops and branches should be collected in a manner that allows the majority of needles to remain in the forest, well spread across the harvesting area. The needles have a high concentration of nutrients and are therefore valuable to the forest's re-growth.

Collecting residues at a Finnish harvesting site.

Primarily, the logging residues are gathered and stored in log piles at the roadside, but an alternative is to store them in piles at the clearing areas for subsequent gathering when the needles have fallen off and the material is dryer. The moisture content is normally around 50 and 55% directly after harvesting but declines when logging residues are stored in the field or at the roadside. Normally, logging residues will be left for some months in the clearings, then after some periods of raining (1 month – 1 year) most of water soluble components (alkalis, chlorides etc.) will separate and drop down to the soil. The residue pile is then covered (only upside) by a light tarpaulin for drying the biomass. Then after one some months the biomass is dry and ready for transport.

The transport lorries can have mobile crusher and the crushing can be done during loading. In the Nordic countries, piles should not be left at the site or at terminals after early autumn, because the moisture content will increase at this time of the year. Wood logging residues can be chipped in the forest directly or at the energy plant. The two most common logging residue systems are chipping at the landing and trucking of chips to energy plants and deliveries of loose logging residues to customer. However, the choice of logging residue procurement system is largely determined by the customer's receiving facilities and fuel requirements.

Residues are sometimes collected and made into bundles. Bundling is generally used to make long distance transport more efficient, however it has also been shown to increase the amount of useable biomass that can be recovered from a harvesting site. Systems for delivering loose or bundled logging residues require that the material is sent to an energy plant or terminal with access to crushing equipment. Deliveries of loose logging residues to the energy plant or terminal are competitive as long as the transport distance is not too far. Logging residues are a voluminous material to transport which results in low payloads on logging residue trucks and higher transport costs per km than for other systems.

Stumps

- Harvesting and Storage: Current stump harvesting methods involve the stumps being lifted vertically upwards by machines. Usually, the stumps are being split in the ground before lifted. They are then shaken to remove stones and dirt, before being stacked in small piles at the harvesting site. They are left at the site to be exposed to the weather, which helps remove additional dirt and sand. They are later collected and transported with forwarders to the landing at the road side, where they are stacked in piles and can dry while waiting for collection. There are two common types of extraction methods, fork type and shear type. The fork type resembles a conventional excavator bucket, but instead of the bucket it has 3–7 long fingers where one finger often is longer than the others to facilitate splitting of the stump before it is lifted. The common design feature for a shear type harvester is a powerful frame with two ripper teeth that are placed under the stump to pull them out of the ground. There is also another type of extraction method called root cutters, but there is currently only one unit with a practical application. This is the rotary stump cutter, which is currently only a prototype and has not yet been released onto the market.

Stump harvester working at a site in central Sweden.

Stump parts are difficult to handle as they are bulky and contaminated by soil and stones. Although the stumps have been split on extraction from the clear cut, the stumps are irregular and difficult to load on trucks at the landing. One way of increasing the payload and reducing transport of contaminants is to coarse-grind the stumps directly on the landing. One of the main advantages of this is more efficient truck transport. Other advantages are decreased ash content in the delivered stump wood and increased thermal value. The ratio between input energy (diesel consumption of the machines) and the energy content of the stump wood is also improved. Calculated for the entire chain, from stump extraction on the clear cut to fuel at the terminal, the proportion of input energy in relation to the energy content in the delivered stump wood was 3.1% for the coarse grinding system and 4.1% for the conventional system with grinding at the terminal/heating plant.

Effects on the Ecosystem

Removal of biological material is affecting the ecosystem. Harvesting logging residues causes nutrient losses to the forest ecosystem because tree's nutrient content is stored mainly in branches and especially in needles. This has negative effects on the fertility of the forest soil. Removing logging residues also causes indirectly some other effects on the growth and ecological condition of the forest. It may change the micro climate and microbe activity, acidify soil and also slow down the mineralization of nitrogen in the humus layer. On the other, hand removing logging residues decreases leaching of minerals such as nitrogen 22– 33% during the first few years after cutting. Removing logging residues has effects also on soil acidification. During the growth of the tree crop natural acidification of the ground takes place when the positive ions such as NH_4^+, Ca^+ and K are bound to the growing stock. These ions will be released back to the ground after death of the tree via decomposing and that compensates the acidity. Removal of the logging residues has been observed to effect about 0.1 pH-unit on the acidity of the forest soil in the final felling stands.

Combustion residues i.e. biomass ash is also recycled to the soil in order to compensate for the removal of mineral nutrients and to maintain the buffering capacity of the soil. However, there is still relatively little information and research data on the leaching of the nutrients from the harvesting and storage areas. If there is a need to fertilize the forest because of the harvesting of energy wood, information is also needed on the leaching of fertilizers. This information would also help to effectively allocate the water protection programs.

Biomass Chips Production

The chipping of the wood used for energy purposes can take place both directly in the forest, at a terminal, or at the end user. For chipping at the harvesting site either mobile chippers or trucks equipped with chippers are used. If chipping is performed at a terminal or at the end user, both mobile and stationary chippers can be used. Mobile chippers are normally hired for short periods to produce enough chips for a specific period of use, i.e. they will be used one day per week to produce enough chips for the rest of the week. This reduces investment cost for the end user.

There are various designs of mobile chippers, which vary according to the material used, volume to be chipped, and the desired product. For small capacities, chipping of forest residues, and mobility, i.e. to chip at the harvesting site, the most commonly used devices are small crushers of various designs. For larger capacities and for chipping roundwood, larger drum chippers and grinders are used. Grinders are used to produce a smaller particle size distribution, and for more contaminated material. Chippers are used when the raw material contains more stem wood and when the particle size should be kept larger.

Due to the high investment and operation costs, stationary chippers are not often used in bioenergy plants. They are only justified when the raw material supply at the plant is expected to include a large amount of roundwood, and when the capacity of the energy plant is high. The wood chips are normally transported directly to energy plants but in some cases they are stored in the forest, at the roadside or in the clearing. If chips shall be stored for more than a few days, they should be of the best possible quality. In most cases, wood chips piles are placed outdoor, but if the moisture content needs to be reduced, or restricted from increasing, it is necessary to place the wood chips under a roof. Wood chips are a suitable growth medium for mould fungi and in a pile of wood chips with high moisture content the fungi grows very fast and soon the entire pile can be infected with mould fungi.

Environmental Impacts of Crushing and Chipping

Crushing and chipping biomass are the most significant noise sources. Noise emissions caused by several different wood chipping and crushing machines have been measured. Measured sound pressure levels varied between 116–130 dB(A), when the main parts of the noise levels were around 120 dB(A) or below. Also wheel loaders and trucks used for biomass loading and transportation create some noise.

Handling, Natural Drying and Other Processes at Growing Site – Cultivated Biomass and Agricultural Residues

Harvesting of Willow

Harvesting takes place in the winter (between November and April), after growing period when, the leaves have fallen and the ground is frozen. Willow is harvested at intervals of 3–4 years and the yield can reach 7–10 oven dried tonnes of willow chips per hectare and year, although the first harvesting is normally smaller. The life span of willow plantation is estimated to be more than 25 years. Willow can be harvested, cut and chipped directly on the fields or as whole shoots.

Storage and Transport of Willow

Normally the willow chips are transported directly to heating plants after harvesting, primarily in bulk transport vehicles, but in some cases the chips are stored in a stack. There are a number of problems associated with the storage of willow chips in a stack. Freshly harvested chips stored in a stack break down faster, due to microbial activity. Willow can also be stored as whole shoots in a pile. The advantage with the storage of whole shoots is that the moisture content is reduced from around 50% to approximately 35% between March and September, corresponding to lower losses, higher density of energy, and better quality of the fuel.

Reed Canary Grass

Harvesting of Reed Canary Grass

Harvesting primarily takes place in the spring when a dry product is received with water content of around 10–15%. The seeds are first harvested in the winter/spring two years after sowing and after that at the same time period year after year. The yield can reach 4–7 oven dried ton per hectare and year. The nutrients and elements that cause problems in the boiler have to a large extent leached out during the winter. Because of that, a small amount of nutrients are removed from the area during harvest. A disadvantage with spring harvesting is that the feasible harvesting period is relatively short. There are two reasons for the restricted period. Firstly the ground has to be dry during harvest to minimize the risk for driving damage from harvesting vehicles, and secondly the harvesting must be done before new green shoots are established. The shoots can otherwise be damaged by the harvesting machinery, which will affect the next harvest yield in a negative way. Shoots may also contaminate the harvest, because of the high water content and the high amount of nutrients. There are two common methods for seizing reed canary grass (for energy purposes) during harvesting and these methods result in rectangular bales and round bales. Shredding the grass in the field is an alternative harvesting method.

Storage and Transport of Reed Canary Grass

Storage stockpiles of bales shall be covered. Since reed canary grass has low water content there is a relatively small risk for microbial activity during storage. Round bales withstand rain showers in the field better than rectangular bales. Reed canary grass is usually transported in bales because the density of reed canary grass chop is very low. When using rectangular bales rather than round bales, it is possible to make 30–50% heavier loads for transportation. Reed canary grass in bales is profitable to transport up to distance of 70–80 km.

Hemp

Harvesting of Hemp

To make full use of the growing season hemp should be sown as early in the spring as is possible. Hemp for energy purposes is best harvested after the leaves have fallen off, which happens after the frost has set in. The nutrients in the leaves are not removed from the area during harvest and can benefit future crops. Frosts also cause the stems to dry and towards spring the steams become very dry (10 to 20%). Hemp that is harvested in the spring can normally be brought in with low water content. There are two methods for seizing hemp (for energy purposes) during harvesting: in one method, the hemp is pressed into rectangular or round bales, in the other method, loose material is chopped into small pieces. Baling offers an effective means to collect, handle and store the material, but it is also a significant cost item in the production chain. The hemps fiber can lead to problems during the harvest; the strong fibers can cause problems with screw feeds and other types of in-feed equipment where it is getting stuck and entangled. If the hemp is harvested in the spring, the moisture content is very low and it is fragile and breaks up easily. A large quantity of small particles is formed and this dust is given off when handling loose stored hemp, which means that workers handling the material must ensure they use appropriate protective equipment to avoid inhaling it.

Storage and Transport of Hemp

Since the crop is only harvested once a year, often in spring when the need for raw material in heating and power plant is low, storage of the harvested material is often required. Bales can be stored indoors or in a covered pile to protect them from rain. Loose material can be stored in covered silos or in tubes. As with other straw material, it is also important to be able to compress hemp to keep down storage and transport costs.

Straw

Harvesting of Straw

When the grain fields are harvested, the straw is left in swaths. The farmer will normally be interested in having the straw removed as soon as possible to be able to start preparing the soil for the following year's crop. When harvesting straw, a baler machine is used and there are two common technologies used when harvesting straw for energy: rectangular bales and round bales. Shredding the straw in the field is an alternative harvesting method, e.g. when straw is used as a raw material for pellets or briquettes. However, the low density and relatively complicated handling make storage and long transport expensive.

There are numerous of different sizes of bales. The average dimensions of "big rectangular bales", which are the most common bales, are 240 cm × 102 cm × 130 cm (L × W × H). The average weight will vary but typically, a bale could weigh 550–650 kg. There has been some interest in producing bales at a height of 90–100 cm, instead of 130 cm; due to better utilization of truck capacity can be obtained. By having 3 layers of 90–100 cm bales instead of 2 layers of 130 cm bales, the average weight of a truck load can be increased by 40–50%. Almost all straw for large scale combustion in Denmark and Sweden is produced and handled as "big rectangular bales".

Storage of Straw

Straw bales wrapped in plastic stored outside in the field.

Straw bales should be transported from field to storage as soon as possible after pressing. If it rains on the bales after pressing, water will penetrate the bales to different depths depending on the type of bales. During storage, there are losses that are caused by microbial activity because the autumn climate can make it difficult to harvest the straw with sufficiently low moisture content. In both Sweden and Denmark, where straw-fired heating plants are common, moisture content higher than 20% is usually not accepted. Uncovered outside storage is by far the cheapest means of storage but the one with most risk attached. Round bales are most tolerant with outdoor storage and more resistant to rain compared with rectangular bales that are sensitive to rain and should therefore be brought undercover as soon as possible after harvesting. There is a machine on the market that can sheet over two rectangular bales on top of each other laid in a long string, called a tube liner. The bales are lifted onto a platform with a front loader and then wrapped in plastic to form a long sausage as more are added. Bales that can be wrapped in this way can have a water content of max 14%.

Transport of Straw

The straw bales can be transported from storage to the energy plant by either tractor trail or by truck. However, tractor trail should only be used with rather short distances between storage and energy plant. As straw has a relatively low calorific value based on volume and weight, the price of straw is sensitive to the distance to the power plant, which results in regional price differentials for straw. Transport cost reductions can be achieved through increasing volume densities, co-handling with other fuels and minimizing steps in the chain.

Grain and Grain Stalks

Storage and Transport of Grain and Grain Stalks

As grains have been, and still are, a very important farming product, there is a well-developed and thought out logistics system for them. Environmental impacts of handling cultivated biomass and agricultural residues at growing site. Environmental aspects of handling energy crops and agricultural residues are in many cases similar to traditional agriculture. There are dust problems associated with harvesting and storage of these kinds of biomasses. Freshly harvested willow chips stored in a stack break down fast due to microbial activity. Reed canary grass on the other hand has low water content why it is a relatively small risk for microbial activity during storage. If hemp is harvested in the spring, the moisture content is very low and it is fragile and breaks up easily. A large quantity of small particles is formed and this dust is given off when handling loose stored hemp, which means that workers handling the material must ensure they use appropriate protective equipment to avoid inhaling it.

Transportation

Biomass can be transported by road, railway or sea, which results in different costs of transport. Transport by road is the most common way of transport at shorter distances. Longer distances often make transport by rail or sea the most cost effective option. However, it is important to point out that the cost of transport often is very dependent on local prerequisites and general transport costs are often difficult to present. Chipper trucks combine chipping and chip transport. Chipper trucks with a switch-body system for chip containers have become more common in recent years. Compared with traditional chipper trucks, their advantages include better utilization of the chipper. The wood is chipped straight into containers which can be transported to the point of end use. The major advantage with transport by railway is that it enables the transportation of biomass harvested from larger distances. In addition, railway transports result in less greenhouse gas emissions as compared with road transports. However, railway transports require terminals for loading and unloading both in sending and receiving sites. The most important cost influencing factors in railway transports based on a calculation model have been observed to be the following: Calculation basis has been the number of deliveries that a train unit makes per a time unit (hour or day):

- Utilization of load capacity.
- Transport distance.
- Terminal handling.
- Conditions for train shunting at terminal points.

The fixed costs are high for railway transports and in order to reduce the proportion of the fixed costs, they must be spread over a large volume.

Environmental Impacts of Transportation

Environmental aspects concerning biomass transport are mainly related to transport efficiency and dusting/littering during transport. Transports should be planned so that the maximum load can be achieved in order to minimize the fuel consumption and CO_2-emissions. Dusting and littering during transport must be prevented for example by covering the load with appropriate means. This will also increase the traffic safety. Special transport equipment for stumps and forest residues has been developed including special solutions to decrease the dusting.

Handling and Storage at Terminals and at the Point of Use

Biomass Logistics Terminals

Biomass can be transported from various harvesting sites to terminal, which are located at a strategically selected site and have good logistics connections. For example that industrial waste heat could be used for fuel drying while having the fuel stored. These biomass logistics terminals are sometimes used to handle, process and temporarily store fuel prior to delivery to the end point of use. A terminal might be established if the biomass growing sites are located far away from the end-point of use. In this case train, ship, or barge transport would be used to transport the fuel from the terminal to the final destination.

The handling and processing technologies and methods used at the terminals are often the same as those used at the harvesting sites or end points of use. Upon arrival the biomass is normally unloaded into temporary storage facilities to wait for the final transportation. Shipping containers can sometimes be used for the temporary storage. Sometimes the biomass may be placed into uncovered storage facilities. In these cases the location of the site and the restrictions regarding noise, dust and water emissions play an important part in this decision.

Pellet storage must always be covered to ensure the pellet quality is maintained and also to prevent the spread of dust around the site. Mobile equipments are most often used for handling the biomass at terminals. These provide most flexibility for moving the biomass inside the terminals and also for unloading and loading the transport vessels. The exact choice of equipment should be based on the types of biomass to be handled and the transport vessels. In large terminals, i.e. those used for loading larger ships, more automated handling systems and equipment are used.

Some terminals are used also for biomass processing, like chipping and crushing of forest residues, stumps and logs. In these cases it is necessary to have chipping or crushing equipment located at the terminal. The choice between using mobile or permanently installed equipments will be based mainly on the long term prospects of the terminal. Permanent installations are used in terminals with long and high utilization rate providing that the investment analysis is justifiable. If the terminal is expected to be in use only for some months or few years, then mobile chipping and crushing equipments are preferable.

Environmental Impacts at Biomass Terminals

Chipping and crushing are operations that generate high levels of noise emissions. Terminals located close to residential areas normally have restrictions regarding noise emissions. In these cases the possibilities to use mobile chippers and crushers outside is limited affecting the choice between mobile and permanently installed equipments.

Biomass Receiving

Weighing

Biomass is most often delivered to the point of use by truck. The advantage of using trucks is that it allows delivery direct from the harvesting site without the need for unloading and reloading en route. However, the cost effectiveness of using trucks decreases after transport distances over 100 to 150 km. Train transport is a more cost effective with longer distances, as much larger volumes of biomass can be transported with a single delivery. The biomass deliveries are usually first weighed to measure the amount delivered. Samples should be taken frequently to check the quality of the fuel delivered and ensure that it confirms with the delivery contract requirements. The end-point-user often pays the biomass supplier based on the quality of the fuel.

The quality requirements are specified in the fuel supply contract, i.e. energy and moisture content, particle size distribution or maximum dimensions, and content of impurities like metal and stones. Modern truck weighbridge systems are operated by the truck driver without leaving the truck cab. An identifying code, card, or electronic key is used to identify the truck and type of biomass being delivered. Dynamic weighing, i.e. where the truck continues moving slowly over the weighbridge during the weighing process, has been used in some installations, however stationary weighing is still more common as it is easier to ensure the accuracy during weighing.

Train weighing systems are based on dynamic weighing. For train weighing RFID identification chips cards are mounted on each carriage. These contain information on what the wagon is carrying, who the supplier is, and the tare weight of the carriage. RFID reading sensors are placed next to the rail track and as the wagon passes over the scales the information in the RFID chip card is transferred to the weighing system database.

Receiving Chipped and Crushed Biomass and Pellets from Trucks

The emptying of trucks delivering already chipped or crushed biomass can be either performed at the rear or side of the truck. Rear emptying of trucks is performed either by tipping the container, or by an in-built conveyor. Similar methods are used in side emptying. Pellets are delivered either in containers or in self-unloading trucks. These trucks will use pressurized air to unload the pellets pneumatically into storage silos. If containers are used, the pellets are unloaded into receiving hoppers similar to that used for receiving chips. Extra care must be taken with dust control and explosion protection when handling pellets due to their low moisture content and dusty nature. Pellets are easily damaged so they should always be stored indoors and should be protected from moisture in order to maintain their fuel quality.

Normally trucks are emptied directly into the receiving hopper. In small plants the receiving system may also act as a storage system, whereas in large plants the receiving system forms an intermediate between trucks and larger storage silos. In cases where dust will not create a problem the receiving hopper can be installed outdoors without any walls or roof. However, it is recommended to cover the hopper to protect biomass from rain and snow ingress into the hopper. Dusty biomass should be emptied indoors to prevent the dust spreading around the plant and to reduce health and safety risks.

The size of the receiving system and the capacity and number of the hoppers must be calculated based on the amount of biomass used by the plant, the truck size and type and the time periods when deliveries are made. To avoid the risk of bridging in the receiving hopper, the hopper geometry should preferably include vertical, or near vertical, walls. Concrete is most widely used for hopper wall construction; however steel can also be used. In wintertime the cause for bridging can be icy biomass and snow with it. In some of those cases fuel silo walls can be equipped with electrical heating systems to avoid bridging.

One type of receiving pocket fitted with chain reclaimer and used for rear unloading trucks.

The reclaiming equipment used inside the hoppers will depend on the hopper size, the biomass handled, and the capacity needed. Where only chips will be handled, screw-type devices can be used. Where the biomass includes forest residues and bark, it is recommended to use either chain reclaimers or hydraulic stokers. In recent years chain reclaimers have been most popular due to the lower investment cost as compared to hydraulic stokers. Leveling rolls, or breaker rolls, should be installed at the discharge end of the hopper to control the flow of material from the hopper into the next conveyor, and to help break up of frozen lumps in the material.

Example of a levelling roll arrangement in a chip receiving hopper.

Receiving Chipped and Crushed Biomass and Pellets from Trains

Train wagons used for carrying chips and crushed biomass are normally designed for unloading at the side. Unloading can be achieved by tipping the wagon or by using front-end loaders to unload the biomass and transfer it to receiving hoppers. Alternatively containers are used and are lifted off the train by mobile equipment before being unloaded by tipping. If pellets are delivered by train they can also be in containers that are lifted off the train for emptying. Alternatively wagons designed for bottom unloading can be used. The equipment and processes used for receiving and handling the biomass and pellets are the same as when receiving from trucks. Similarly, the same environmental and safety issues apply and should be controlled in the same way.

Receiving Logs, Chipped and Crushed Biomass and Pellets from Ships

When biomass needs to be transported for long distances, or between countries, ships or barges are often used. Biomass is transported in many different sizes of vessels, from barges carrying a few thousand ton to Panamax vessels carrying fifty-thousand ton or more. Loading of crushed and chipped biomass and pellets into ships and barges is carried out using various types of conveyors, but most often belt conveyors. Loading and unloading of logs is carried out using pier mounted cranes and mobile excavators equipped with special log grabs. Unloading of the crushed and chipped biomass and pellets is performed by various means.

Some vessels are self-unloading and have a built in conveyor systems that can transfer the biomass to receiving hoppers located on the pier. Other vessels without self-unloading systems are most commonly unloaded using pier mounted harbor cranes fitted with large bucket grabs. Sometimes large mobile excavators equipped with bucket grabs are located on the pier and used for unloading smaller ships and barges. Self unloading ships can more easily control dust emissions and spillages when compared to the use of cranes and excavators. Receiving hoppers should preferably be equipped with dust collection devices to minimize dust emissions.

Mobile excavator unloading chips from a ship in Finland.

A very important safety issue in unloading crushed and chipped biomass and biomass pellets from ships concerns the risk of personal injury or death from entering confined spaces where there is a lack of oxygen and high levels of poisonous gases. There have been several deaths from carbon monoxide poisoning in cargo holds during biomass unloading. During transport, which sometimes can last several weeks, natural biological activity in the biomass causes a build-up of dangerous gases in the cargo holds. Therefore it is critical that the cargo holds are properly ventilated and measurements of carbon monoxide and oxygen levels taken prior to anyone entering the holds for unloading or cleaning operations.

The unloaded biomass is placed in hoppers on the pier from where it is conveyed further with belt conveyors. Pellets should be placed directly into covered storage; however crushed and chipped biomass should be first screened to remove oversize pieces and impurities. When planning ship receiving systems for biomass it is important to allow enough storage capacity so that the ships can be unloaded quickly and do not need to spend excessive amounts of time in port, which is expensive and increases fuel costs. Further handling of the biomass is similar to the methods used at the end user facilities.

Receiving of Logs and Biomass that Needs Crushing or Chipping

When logs, or biomass that needs chipping or crushing, is received they are either unloaded directly into the crushing or chipping systems, or placed into outdoor storage. Some biomass, like branches and stumps, can be delivered in containers or self-emptying trucks designed for unloading at the rear. In this case these trucks can unload directly onto a special receiving deck fitted with a conveyor that will meter the flow into a crusher feed conveyor, or directly into the crusher. These types of receiving decks are designed so that the truck can reverse onto the deck for unloading. Trucks fitted with cranes can unload themselves, but otherwise mobile or stationary cranes are used. Often rubber wheel front-end loaders fitted with special grabs designed for lifting logs or loose biomass are used. Wheeled excavators fitted with similar grabs are also used when greater lifting height or reach is required. Stationary knuckle-boom cranes can be used however they are quite expensive and therefore can only be justified when the volumes of fuel to be handled is high.

Chipping and Crushing of Logs and Larger Biomass Pieces

Logs used for biomass production are normally chipped with the bark still on the log. The logs are metered onto a chipper feed conveyor line, which can be a belt or chain conveyor, where they will pass through one or more stone traps and a metal detector before reaching the chipper. The loading of logs onto the log deck and the further conveying of logs into the process will produce quite high noise levels. Therefore care should be taken in planning the layout and equipment design in order to limit the noise levels at sensitive areas. The lifetime chipper knifes and wear-parts is very much affected by impurities such as stones and metal, and serious damage may occur to the chipper if metal passes into it.

Therefore the use of metal detectors is very important. It is also critical to note that there must be well designed safety systems. Strict safety procedures must be established in the removing process of biomass from the chipper and crusher feed conveyors as it sometimes requires the operator to enter the feeding conveyor while the chipper or crusher is still running. Any mistakes or accidents while the operator is on the feeding conveyor can easily result in fatal injury. The main points when selecting the chipper design should be production capacity, the size of logs to be chipped, investment and operating costs and safety of the design.

When crushing larger biomass pieces like stumps and branches, there are two main alternatives. Since the early 1990's the most commonly used type of crusher for this bulky biomass is a large, vertically fed and slowly rotating crusher. The rotation speed is normally in the range of 30 to 100 rpm and the capacity ranges to about 200 loose m^3/h. The investment and maintenance costs for this type of crusher are quite high. The annual production capacity is therefore an important factor in determining whether or not such an investment is justified. Biomass crushers used in modern applications are less sensitive to metal and stones when compared to chippers. Small metal pieces will cause only minor damage and wear. However, larger pieces can still cause major damage to the crusher. Therefore the use of metal detectors is recommended in order to prevent large metal pieces entering the crusher.

From the late 1990's several suppliers have developed large, horizontally-fed, biomass shredders designed for handling bulky biomass like branches, stumps, and even bundles of biomass. For these types of crushers the rotation speed is in the range of 500 to 1,000 rpm and capacities range is to about 400 loose m^3/h. As these crushers operate at a much higher speed than the slow speed crushers, it is more important to prevent large metal pieces entering the crusher. There is also a higher risk of creating sparks which may cause explosions, so suitable protection methods should be adopted if the biomass is expected to have explosive properties.

Biomass Conveying to Screening

After receiving, crushing, or chipping, the biomass is transferred to screening. Several types of conveyors may be used The choice of conveyor type is mainly dependent on the layout of the plant, but also the type of biomass and the capacities needed play an important part. In Nordic countries the most commonly used conveyor type is the so called scraper chain conveyor. The main advantage of this conveyor is that it is dust tight and able to convey at steep inclines. In the handling of dusty fuels that can be explosive it is important to include explosion pressure relief doors on chain conveyors. Belt conveyors are a more economical choice, especially when conveying distances are greater than about 50 m. However, there are disadvantages because of the fact that they require more space as their incline is more restricted. In order to prevent problems with biomass slipping back on the belt the maximum recommended incline in Nordic environments is 15 degrees and it is preferred to keep the incline below 12 degrees if possible.

Another disadvantage with belt conveyors is that they are not dust tight. In the handling of dusty biomass extra care must be taken to minimize dust cloud generation and to collect dust at the points where it is generated. This requires the use of self-cleaning dust collection cartridge filters, or similar. The dust is deposited back onto the conveyor so that it is conveyed further along the process. There have been some experiments with biomass dust suppression using atomized water mist sprays, also known as dry-fog technology.

The use of this technology is not yet widespread in biomass handling systems. It is important to consider the maximum lump size and impurities when selecting the conveyor type to be used between the receiving pockets and the screening system. Large frozen lumps or large stones t may be present in the biomass flow. Screw conveyors are not recommended for use between receiving and screening. Screw conveyors should only be used in this phase if the handled biomass is certain to be free of large lumps and impurities, the final quality fulfills boiler requirements.

Biomass Screening and Reject Crushing

It is recommended to always screen green biomass after it has been received as it is normal for oversize pieces to be present in the flow. It is also common for stones and metal to be present, in varying amounts depending on the type of biomass and where it has come from. Logs and whole tree chips can contain metal pieces embedded in the wood during the tree growth, i.e. nails, bullets, etc. Screening systems in Nordic installations are normally located indoors to protect them from unfavorable weather conditions and also to minimize noise emissions. Commonly disc screen technology is used to remove oversized particles and reject them either to a crusher, or to a reject container.

Often a short belt conveyor is used to spread out the flow across the width of the screen, to provide bypassing of the screen and crusher, and also to allow a magnet to be used to remove metal pieces. If the conveyor between the receiving pocket and the screening system is a belt conveyor, it is possible to place the magnet at the discharge of this conveyor and dispense with the shorter belt conveyor. In this case a separate bypass chute should be provided to allow the screening system or just the crusher to be bypassed during maintenance. Both self-cleaning belt magnets and non-self cleaning magnets may be used. The metal collected should be deposited and located into a container for recycling.

There are two alternatives for handling the screen rejects or oversize pieces. If the amount of rejects is expected to be small, it can be economically attractive to discard the rejects into a container to be treated either as waste or sold to another user with the ability to receive them. Alternatively the rejects can be collected and in case of large volumes be crushed by renting a suitable crusher for that purpose. Where the amount of rejects is expected to be quite large, it is normal practice to process them directly with a crusher, and to feed them back into the biomass flow. There are many types of crushers that can be used to this purpose, ranging from high speed hammer mills to simple slow speed twin-rotor crushers. The slow speed crushers typically operate at about 20 to 50 rpm and have capacities of to about 200 loose m^3/hr depending on the material being crushed. The key differences between hammer mills and heavy-duty twin-rotor crushers are shown in table.

Table: Main differences between crusher designs.

	High-speed hammermill	Heavy-duty twin-rotor slow speed crusher
Rotation speed	>500rpm	<100 rpm
Maximum lump size of product infeed	Approx. <400mm	Approx. <600mm
Quality of fuel produced	Smaller size distribution	Larger size distribution
Risk of explosions when handling dry dusty fuel	Medium	Low
Risk of explosions when handling green biomass	Low	Low
Maintenance costs	Medium to High	Medium
Investment cost	Medium	High
Investment cost vs. capacity	Low to Medium	Medium to High

Biomass Conveying to Storage

After screening and crushing the biomass should be conveyed to an intermediate storage. Similar selection criteria should be used as for the conveyors between receiving and screening. At this phase in the process the biomass properties are quite consistent and this allows the use of other types of conveyors such as screw conveyors, pipe-belt conveyors, and air-supported belt conveyors. Pipe-belt conveyors are more expensive than normal belt conveyors and they can make turns, provided that the conveying route is long enough. This helps to eliminate the need for a transfer point and second conveyor in more complex layouts.

Pipe-belt conveyors are also fully enclosed over the conveying distance leaving loading and discharge points open and enabling easier odor control. Air-supported conveyors have become popular in recent years. These use a cushion of low pressure air to support the belt inside a steel pipe that is part of the conveyor gallery structural frame. This minimizes the need of moving parts such as idler rollers, and therefore also minimizes maintenance needs.

Where site space is limited, and a compact layout is necessary, belt bucket elevators may be used. Their design must be made carefully. Bucket elevators are more sensitive than other conveyors to the presence of impurities, large lumps, and variations in flow rate. Mistakes may lead to blockages and mechanical failure. The maintenance needs of bucket elevators are higher than for other conveyors and this should be considered when selecting the type of conveyor to be used. The number of storage silos will directly affect the design of the conveyor system. When more than one silo is used, a distributing conveyor must be located over the silos to allow the flow to be directed to each silo.

When designing the layout of the conveyors and other plant items careful attention must be made to the need for maintenance access and providing platforms and stairs to allow this. The location and number of fire escape routes must also be considered. Regulations in each country dictate the maximum distances allowed between escape routes. It is recommended to avoid the use of spiral and helix staircases, even if they are not forbidden by regulations, as they make maintenance access more difficult and if there is an accident it can be impossible to carry a stretcher on the stairs.

Storage and Reclaiming of Prepared Biomass

Fuel storing capacity at the point of use is important; a recommendation is to have storage for at least 2–3 days operation. In case of difficult traffic or bad weather conditions the transportation and fuel flow can be stopped. In order to protect the fuel from unfavorable weather conditions it is preferred to store prepared biomass under a roof, so that the moisture content does not rise. Placing the biomass inside a silo provides the highest level of weather protection and also prevents dust emissions to the surrounding areas. The spreading of dust and chips is a significant problem in uncovered piles located in windy areas.

Two types of silo have been most commonly used in the Nordic countries for loose biomass. The most common is the circular silo equipped with a cantilevered radial screw reclaimer or similar. These are commonly used for storage capacities up to 10,000 loose m^3 net, although most of the existing installations are below 5,000 loose m^3 net. The other silo alternative used is the so-called "A-Frame" storage equipped with linear screw reclaimers or similar devices. A-Frame silos are often used where larger storage volumes are required. In some cases, due to layout restrictions, or due to the quality of the fuels handled, multiple circular silos will be used in preference to an A-Frame silo.

Two large circular silos for biomass chip storage.

In modern installations circular silos are most commonly made from pre-fabricated flat concrete wall sections, similar to those used for large water tank construction. These are placed on top of a concrete base that may be constructed above or at ground level. The ratio of maximum fuel height to silo diameter has an important impact on the ability of the biomass to flow freely or bridge inside the silo. On larger circular silos, approximately ≥ 22m in diameter, a steel frame roof covered with insulated industrial roofing panels is fitted on top of the walls. In smaller silos a welded steel plate roof is often used. Access stairs should be provided to the top and centre of the roof so that the feeding conveyor and instrumentation devices can be maintained. Explosion pressure relief doors should be placed in the silo roof when the biomass is expected to have explosive properties.

A-Frame silos have a concrete foundation upon which a steel frame roof is erected and covered with insulated industrial roofing panels. The floor of the silo can be located at ground level, but in that case the reclaim conveyor must be placed in a tunnel below ground level. If the silo floor is raised above ground level the need for tunnels can be avoided. The feeding conveyor from the screening system should transfer the fuel to the top and centre of the silo.

Inside the A-Frame silo a shuttle conveyor is located under the roof and is used to distribute the fuel along the length of the silo. Various silo arrangements are possible, and the layout can be chosen according to the site layout and capacities needed. The decision on whether or not to place the silo floor at ground level or raised above ground level should be made on a project-by-project basis as the local ground conditions and civil works costs have a big affect on this decision.

A-Frame and circular silos for chip and bark storage at a pulp mill.

Both A-Frame and circular silos should include a large access door that allows the fuel to be emptied with an excavator or front-end loader in the event of equipment breakage. The reclaimer used inside circular and A-Frame silos is most often a screw reclaimer. Screw reclaimers are able to reclaim sticky and nonfree-flowing materials, so are preferred when forest residues and bark are handled. When quite good quality chips, sawdust, or similar freeflowing materials are handled, it is be possible to use other designs of reclaimers.

It is important to select a reliable and well proven reclaimer design in order to minimize the risk of unexpected breakages and blockages. The operation of this equipment has a direct impact on the boiler availability, so a high level of reliability and up-time is required. The reclaimer maintenance intervals differ greatly dependant on the fuels handled and the materials used for the equipment construction.

Conveying to the Boiler Silo

After storage the biomass should be conveyed to the boiler silo. Similar selection criteria should be used as for the conveyors between screening and storage. If multiple silos are used, the conveyor system should be designed to mix the biomass from the different silos. Control of dust is again an important issue especially as the boiler transfer conveyor normally needs to rise high up into the air and any dust spillages can otherwise spread over a wide area.

Handling of Straw at the Plant

Example of straw handling at the plant.

At the small power plants, the straw is mostly unloaded with a fork-lift truck. At the larger plants, the unloading is carried out with a so-called overhead travelling crane. The crane is capable of unloading the bales from the truck as well as it also registers the weight of the straw and its moisture content. However, at small plants, the moisture content is registered with a spear which is manually pushed into each straw bale. Generally, the storage capacity of power plants is large enough for only a few days. A great deal of dust is generated when breaking up and roughly chopping straw bale, which must be dealt with suitable ventilation equipment.

Control of Emissions to Air and Water

Self-cleaning dust collection filters are used to capture dust generated during unloading and handling processes. The dust is returned to the main biomass flow in the conveyors. When unloading dusty biomasses drivers should be able to operate the system from a location where they are not exposed to the dust. For limiting dust emissions water dispersion systems can be used. If biomass is placed into outdoor storage care must be taken with control of emissions to air and water. Chips and dusty materials can be easily blown around the site. These solids can easily pass into rainwater drainage systems where they should be collected in solids separators before the drainage water is released to any local bodies of water.

Other harmful emissions like oil leakages can occur due to mechanical failure of the mobile equipment and trucks operating in the storage area. Therefore rainwater handling systems should also include oil separation devices. Rubber wheel loaders and similar mobile equipment have high noise emission levels, in the range of 108 dB(A), so their operating times are often restricted at sites close to residential areas. This should be considered when applying for permits, and also during layout planning of the plant.

Both chippers and high speed shredders emit high noise levels. Chippers can produce noise levels of approximately 110 dB(A). It is therefore necessary to enclose chippers and crushers inside buildings designed to enclose and suppress the noise emissions. Care should be taken with the design of the feeding conveyors and how they enter the buildings to ensure noise emissions through these openings are limited. Slow speed crushers often use hydraulic drive systems and these can also be quite noisy, so care should be taken by placing the hydraulic power unit inside a noise enclosure, a room, or an underground area.

Biomass Dewatering and Drying at Industrial Scale

The fresh deforested trees have a moisture content of about 50%. The moisture content is depending on the season, type of wood and part of the tree. There are several means of reducing the moisture content by natural drying. In different production processes there is, however, a need to dry the material by active heat transfer. This is the case for saw mill products and pellets. For plants with only district heating purpose there is normally no economical or technical interest in drying wet biomass before the intake (some cases exist, preferable where waste heat can be used). In the pulp industry debarking of the stem wood is done at the mill and the bark is used as a fuel for the heat production. In the cold winter period with freezing conditions a thawing of the stem wood with steam is leading to significant increase in moisture content. This will of course aggravate the combustion. A common technique to reduce moisture content is wet pressing (mechanical), after a crushing, where the moisture content is reduced to below 60%. At the same mill a second drying step can be installed with an active heat drying. The technique can be of either bed, drum or stream drying methods.

Environmental Aspects

- Emission to air: Volatile Organic Compounds, VOC, emissions can occur during storage and drying of biomass. Emission of monoterpenes from bark has been determined to be high immediately after building a pile of fresh bark but declined within few days to acceptable levels. In the drying process of bark emission of terpenes, alcohols, fatty acids and resin acids have been detected.

Laboratory tests have shown that during drying of forest chips and sawdust VOC compounds that will release are terpenes, alcohols, aldehydes and carboxylic acid. Emissions of aldehydes and terpenes have also been found in bulk storages of wood pellets. Typical compounds are hexanal, pentanal and monoterpenes.

- Dust emissions: There is always some dust present when handling especially dry fuels produced from biomass. Dust is primarily an occupational health risk to the plant's employees. The operation of the plant should be organized in such a way that dust emissions are not spread to the surrounding air.
- Noise emissions: Crushing and chipping biomass are the most significant noise sources. Noise emissions caused by several different wood chipping and crushing machines have been measured. Measured sound pressure levels varied between 116–130 dB(A), when the main part of the noise levels were around 120 dB(A) or below. Also wheel loaders and trucks used for biomass loading and transportation create noise emissions.
- Emissions to water: Leachages from biomass outdoor storages can lead to solid organic and dissolved organic emissions into watercourse. Toxicity of leachates of pulp mill bark, saw mill residues and chipped residues from forest loggings in the laboratory scale has been examined with toxicity test. According to the results leachates from bark and saw mill residues were acutely toxic to the test organism Daphnia magna. The toxicity increased with time to a certain extent. The toxicity was caused partly by phenolic compounds. Leachates of the chipped logging residues showed no toxicity.

In laboratory storage tests carried out by lysimeters the quality of leachates from forest chips piles (green and brown) and fresh sawdust piles were determined. The test results showed that COD of the leachates varied between 2,000–9,000 mg/l, TOC 900–1,400 mg/l and pH 4–5.6. Total COD concentration of the leachates was 0.1–0.5 weight-% and TOC 0.1–0.2 weight-% of dry weight of the material. Highest amount of organic material was dissolved from green wood chips.

Biomass Boilers and their Feed Systems

Biomass boilers differ from oil or gas-fired boiler counterparts in a number of ways. In particular:

- Boiler size: Biomass boilers are physically larger than oil and gas-fired equivalents. More space surrounding the biomass boiler is needed (access for cleaning, fuel-feed and ash-removal systems etc).
- Boiler responsiveness: Biomass boilers are slower to respond to changes in heat demand. This can be managed by intelligent control, thermal storage and/or the use of top-up conventional boiler capacity. The turndown ratio 20 of biomass

use of top-up conventional boiler capacity. The turndown ratio 20 of biomass boilers is lower than that of gas or oil-fired boilers.

- Fuel storage: More space is needed on site for fuel storage. Boilers that are fired on oil or liquefied petroleum gas (LPG) need space for fuel storage, but biomass systems require far more space due to the lower energy density of the fuel.
- Fuel delivery: Due to the lower energy density of biomass fuels, access suitable for large vehicles and/or deliveries that are more frequent is needed.
- Emissions: It is more difficult to control emissions from biomass boilers than from oil and gas-fired boilers. Biomass boilers produce fly ash and particulate matter that may require the installation of additional abatement equipment. In addition, nitrogen oxide (NO_x) emissions per unit of heat generated are usually higher than those from fossil fuel-fired boilers. If NO_x emissions are not sufficiently managed by the control of combustion conditions, then additional flue-gas abatement equipment may be required.

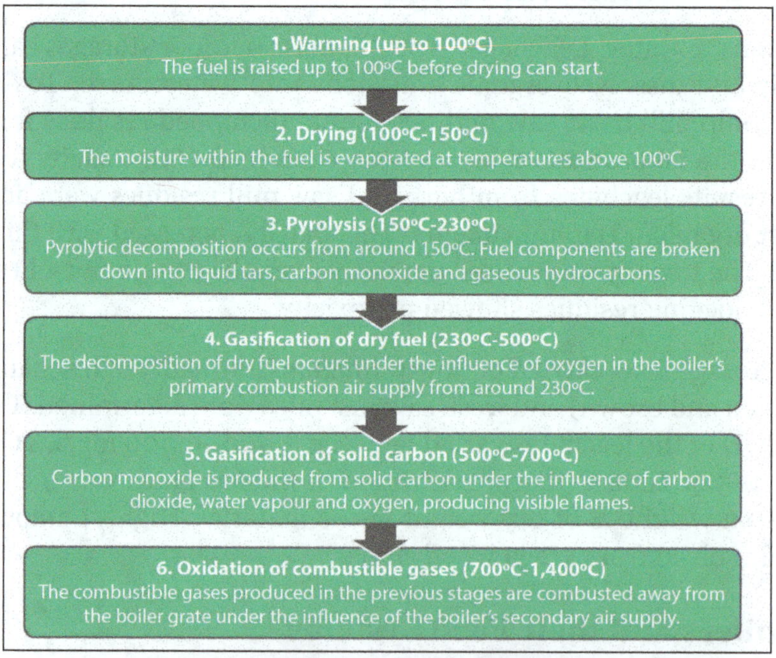

The six stages of biomass combustion.

Boiler Types and Characteristics

There are various factors that determine biomass boiler selection, including:

- Heat generating capacity required.
- Fuel type and characteristics.
- Required boiler responsiveness and turndown ratio.

- Requirements for limiting emissions.

There is a wide range of biomass boiler types available that can meet end users' specific preferences and requirements. This guide describes the most common types of boiler available. Additional advice can be sought from biomass boiler suppliers or manufacturers. Biomass boilers can be classified by a combination of the following parameters:

- Boiler Capacity/rating kW.
- Automatic versus manual feed of fuel.
- Stoker type (fuel supply).
- Ignition: automatic ignition versus slumber mode.
- Fire tube orientation.
- Fuel type (s).

Boiler Capacity

The majority of non-domestic biomass boilers have a heat output range of around 50 kW to 5 MW and are designed to produce heat in the form of hot water. Boilers of capacity greater than 1.5 MW can generate steam for heat transfer. Boilers that use thermal oil as the heat transfer medium are usually available above 3 MW.

Apart from those using logs, most non-domestic boilers are automatically fed with fuel. Smaller boilers may only have a small integral fuel hopper that requires regular manual reloading with wood pellets. Some larger boilers, for certain fuel types such as straw may also be manually fed, particularly at sites where there is lowcost manual labour available. Automated systems are more complicated and less mechanically robust than manual systems.

Automatic Stoker Types

Automatic stokers introduce the fuel into the boiler for combustion and are an integral part of the boiler design. The stoker design will be determined by factors including the boiler capacity, the type of fuel and its moisture content (MC), and the degree of control needed for combustion efficiency and boiler emissions. Stokers are either overfed or underfed, introducing the fuel from above in overfed stokers and below for underfed stokers.

Stoker Burner Boilers

The stoker burner is an overfed stoker type. They are more common for use with smaller capacity boilers (typically 30 kW to 50 kW), but can be used for boilers up to 1 MW.

Using an auger, fuel is transported onto the grate, which can be in the form of a cast iron tube. Stoker burners are suitable for wood chip and wood pellets, provided the MC is less than about 30%. The small combustion chamber size means that there is limited refractory material present to reflect heat onto and drive moisture from the fuel. This makes this boiler type more sensitive to MC, but more responsive to changes in heat demand than other boiler systems. Ash is pushed off the grate into the ash pan by the auger when introducing new fuel. Primary and secondary combustion air is normally provided by a single fan, which restricts the ability to limit the temperature at the grate while maintaining a higher temperature in the secondary combustion zone.

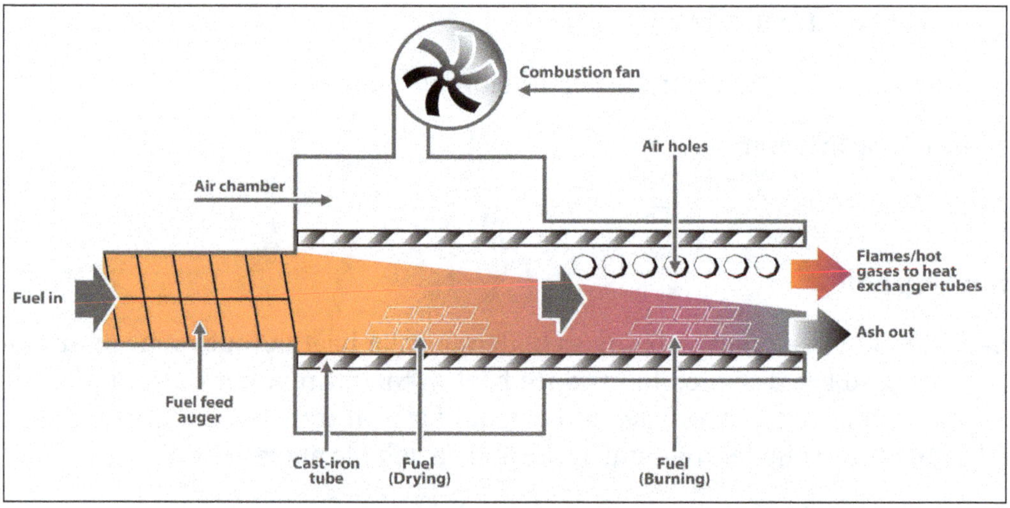

Schematic of a stoker burner.

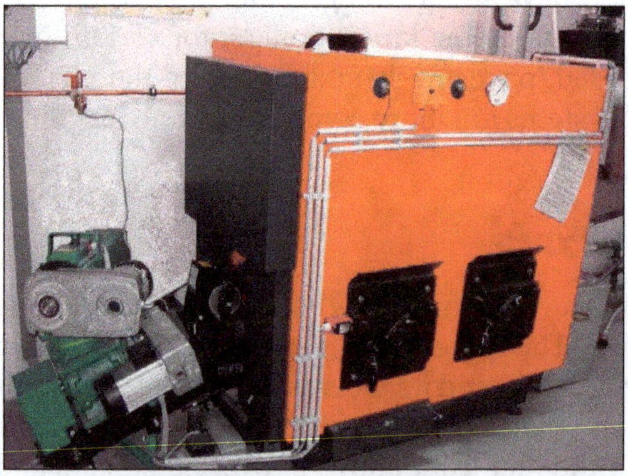

Small stoker burner boiler.

- Advantages of Stoker Burner Boilers are:
  - Simplest type of automatically fed boiler.
  - Relatively low cost. Good turndown ratios with low fuel consumption in slumber mode (if used).

- ◦ Relatively rapid response to changes in heat demand.
- Disadvantages: Boiler performance is very sensitive to changes in MC and chip/pellet size. If the MC is too high, black smoke is likely to be produced due to incomplete combustion. Smaller systems may not have separate primary and secondary combustion air fans. This makes it difficult to control the temperature of combustion and may result in ash melting (fusing) on the grate, and can also make it difficult to control the formation of NO_x. The first problem can be prevented by incorporating a water cooling circuit in the grate and the second by recirculation of flue gases into the combustion chamber.

Moving-grate Boilers

Moving-grate boilers (also known as stepped or inclined grate boilers) are overfed and are available in a wide capacity range (10kW to 10MW), and offer great flexibility for boiler design and fuel choice. Fuel is introduced onto a series of inclined flat panels (also called fire bars) which move in sequence, causing the fuel to shuffle down the incline towards the ash pan. This sequence of fuel movement means that the combustion stages occur at different places on the grate, affording greater control of the overall combustion process. By changing the speed of the grate, the duration spent in the drying zone can be changed, ensuring that fuels with a wide range of MC can be accepted by the boiler. This helps complete combustion, preventing black smoke (soot).

As well as being able to accept fuels with a range of MC, these boilers can accept fuel with high absolute MC. The space inside the combustion chamber allows for high levels of refractory lining, which means that the capacity to reflect heat onto incoming fuel is significant, allowing fuels of high MC to be accepted. On the downside, the high levels of refractory lining mean that the response of the boiler to changes in heat demand is slow.

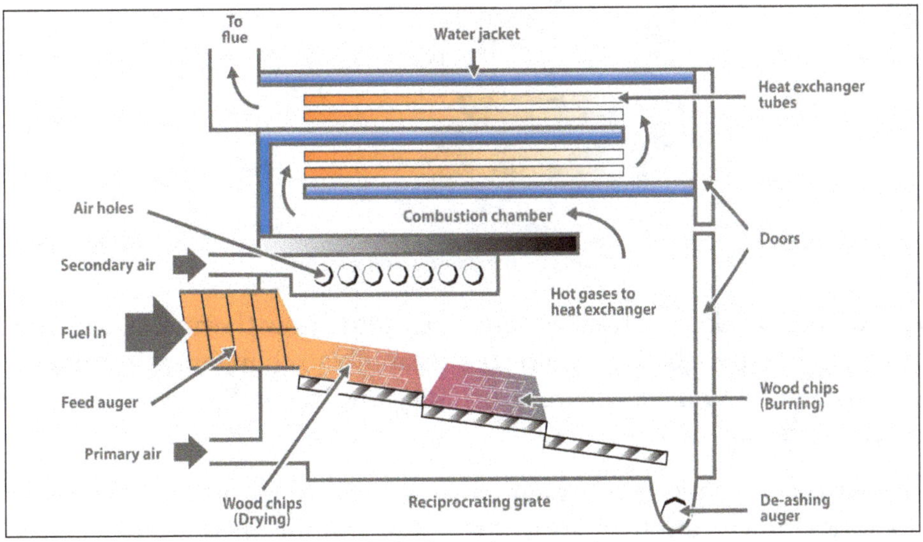

Schematic of a boiler using a moving grate.

Moving-grate boilers can burn wood chips and pellets. However, using chips is more common because they are cheaper and the MC tends to be higher than that of wood pellets. Fuel is introduced using either an auger or a ram stoker. Separate fans for primary and secondary combustion air provide additional control.

- Advantages: Designs can accept fuels with a range of MC (up to 55%). The moving grate means that problems with clinker (slag) forming on the grate are less likely to occur, which means that these boilers can be fuelled with a wider range of biomass compositions. Separate primary and secondary combustion air fans enable a high level of control.
- Disadvantages: When designed for high MC fuels, the additional refractory material means that response to changes in heat demand can be slow and the fuel consumption in slumber mode is high.

Underfed Stoker Boilers

In the underfed stoker, fuel is fed by an auger to the combustion chamber from under the grate. The fuel is forced up through an inverted cone to form a dome of fuel on the grate where combustion takes place. As more fuel is introduced, the ash falls from the periphery of the dome into an ash bin below. Wood chip and pellet fuel can be used, usually with a MC less than 30%. Underfed stoker boilers are available up to around 5MW.

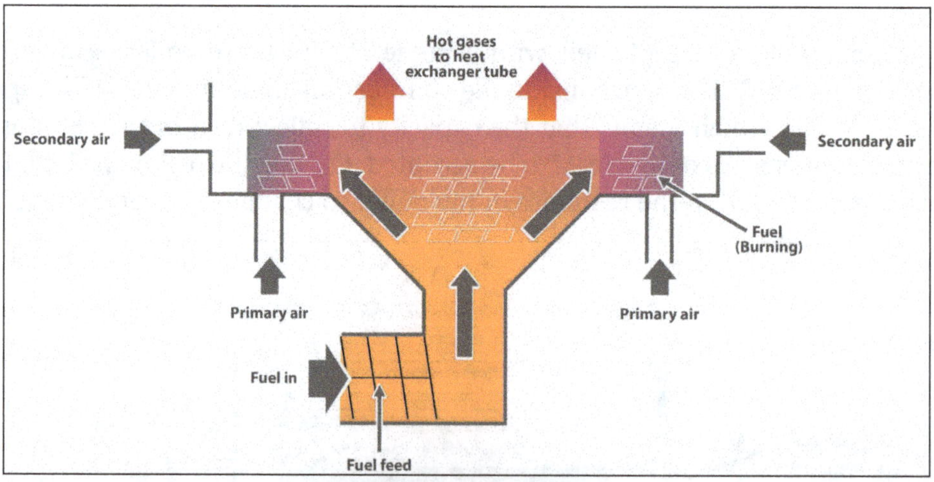

Schematic of an underfed stoker.

Like the stoker burner design, the low refractory mass (low thermal mass) and low fuel inventory mean that these boilers can respond rapidly to changes in heat demand. Most underfed stoker designs have separate primary and secondary combustion air fans, aiding complete combustion.

- Advantages: Simple mechanism with a wide range of boiler sizes available. Good turndown ratios with low fuel consumption in slumber mode (if used). Relatively rapid response to changes in heat demand.

- Disadvantages: Boilers can only accept chips/pellets within a narrow range of MC and particle size.

Underfed boiler combustion chamber.

Other Boiler Types

Log Boilers

Log boilers are manually batch-fed and can burn wood logs or briquettes made from forestry waste or brash. Log boilers are relatively cheap, but operation requires more manual labour. Consequently, they tend to be used at sites where there is already labour available, such as workshops and farms. Log boilers are most commonly used in smaller scale systems. An accumulator tank, holding at least 50 litres of water per kW of boiler rating, should normally be connected to a log boiler. Logs need to have a MC of typically less than around 20% throughout to avoid soot due to incomplete combustion.

- Advantages: Relatively low cost, Simple and robust, easy to maintain and fuel is generally low cost.
- Disadvantages: The fuel must be relatively dry, typically less than 20% MC. Requires labour and frequent intervention.

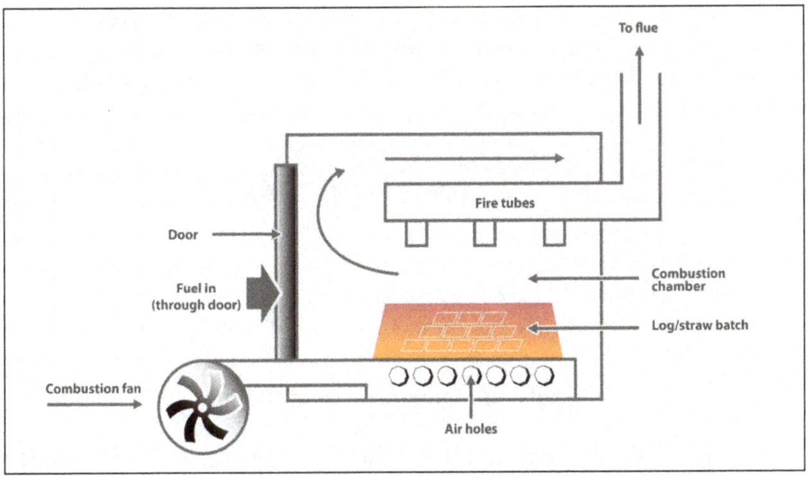

Schematic of a batch boiler.

Straw Boilers

500 kW straw boiler.

Normally, the combustion chamber of straw boilers is manually batch-fed with one or more bales at a time. Straw boilers are generally designed for standard bale sizes. The pros and cons of straw boilers are similar to log boilers. However, because straw is very light, it can be difficult to control emissions, necessitating measures such as additional cyclones and electrostatic precipitators. Straw often contains high concentrations of chlorides, which promotes slag formation by reducing the ash fusion temperature. Straw boilers are specifically designed to manage this.

Fluidised-bed Combustion of Poultry Litter

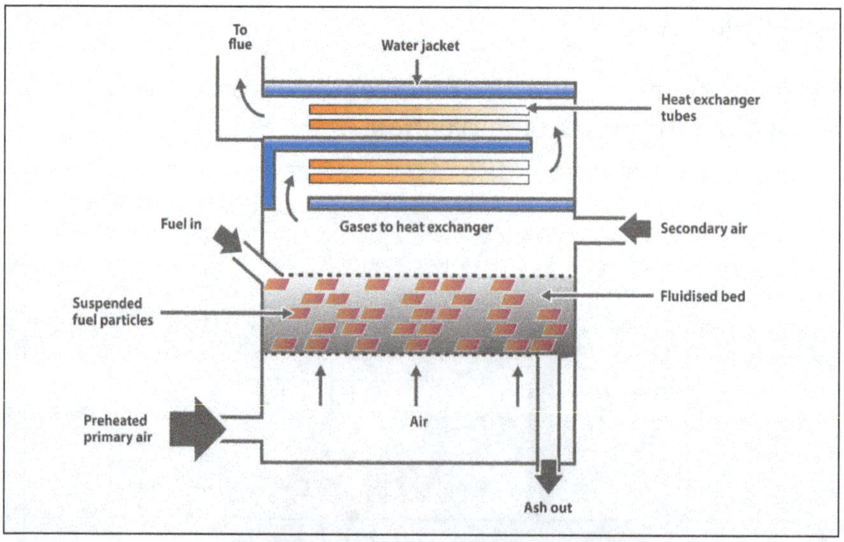

Schematic of fluidized-bed combustion.

The combustion of poultry litter is usually, though not exclusively, carried out in fluidised bed combustion (FBC) boilers. The technology was originally limited to power station-scale plant, but now boilers down to around 500kW are available for use on farms. The principle of FBC is that the fuel particles are suspended and combusted in a bed of hot sand (or similar mineral), which is first heated by an oil or gas burner and fluidised by jets of air from beneath. This provides continuous mixing of the fuel with the primary air and thus more effective combustion, particularly for lower quality and variable MC fuels. Secondary combustion air is introduced above the fluidised bed to ensure complete combustion.

Types of Fuel Feed

For boilers where the fuel is fed automatically, there are two main types of feed system.

- Auger feed: An auger is a screw mechanism which, when rotating, transfers fuel material along its length. A big advantage is that fuel can be transported around corners and between different vertical levels if more than one auger is used. This gives more flexibility when positioning the boiler and the fuel store. Auger-fed systems are also cheaper than systems based on hydraulic rams. However, auger-fed systems only work well for fuels within a relatively ridged specification of particle size, shape and MC. Significant deviations from the norm on these properties can lead to blockages.

- Hydraulic ram: Compared with auger-fed systems, hydraulic ram systems are more reliable for larger boilers, as they can handle fuels with a range of particle sizes and MC. However, they are more expensive than auger-based systems and cannot operate across more than one level. There is also a small fire risk from the hydraulic oil.

Types of Ignition

Manual Ignition

When the MC of the fuel is relatively high (greater than 45%), the boiler will normally be ignited manually using a gas lance (or firelighters or paper), which is pushed into the firebed until combustion is self-sustaining. Boilers burning fuel with a MC greater than 45% usually have a moving grate to aid ignition. Most moving-grate boilers operate in what is known as 'slumber mode'. This means that when the boiler has produced sufficient heat to meet demand, the firebed is maintained at a very low level by reducing the fuel feed to the minimum required to keep the firebed alight. In slumber mode, the primary and secondary air supply is turned right down or may even be switched off.

- Advantages: Because the firebed is already established, these boilers can be more responsive when heat demand increases.

- Disadvantages: If the boiler is oversized (and other aspects of the system are not well designed), it will be more often in slumber mode, reducing the overall efficiency. Using thermal storage and back-up fossil-fuel fired boilers can help to reduce peaks and troughs in heat demands.

Automatic Ignition

Automatic ignition systems are used in boilers burning relatively dry fuels (≤45% MC). The ignition system fires as required to meet the load demand (normally the demand for heat from thermal storage) and the firebed is not maintained at times of zero load on the boiler. Firing takes place until the firebox reaches the minimum temperature required for combustion to be self-sustaining. The type of system used depends on the size of the boiler. Larger boilers will use gas, LPG or oil burners fitted at the side of the firebox. Smaller boilers will use electric hot air blowers.

Automatic ignition must be compatible with the MC of the fuel. If a higher than normal (≥45%) MC fuel is used, and then the ignition system may not achieve self-sustaining combustion, but rather a smouldering (pyrolysing) fuel on the grate producing carbon monoxide (CO) and hydrogen. These gases must be purged from the combustion chamber before ignition is attempted again, otherwise an explosion is likely. It is strongly advised that boilers with automatic ignition are not used with fuel of a higher MC than specified by the manufacturer.

Boiler Fire Tube Orientation

The heat of the biomass combustion pass through fire tubes, heating the water surrounding them. There are two possible orientations of fire tube i.e. vertical and horizontal.

Vertical Fire Tube

These are found in all three main boiler types (stoker burner, underfed stoker and moving grate). They tend to be found on boilers with capacity up to 1MW, and mostly in the capacity range up to 200kW. The main issue with vertical fire tube boilers is that sufficient clearance must be left above the boiler to allow access for cleaning and maintenance.

Horizontal Fire Tube

Also found in all three main boiler types but are more common on larger capacity boilers. While less vertical headroom is required than that for the vertical fire-tube boiler, greater horizontal clearance is required for access to clean the fire tubes. Also, horizontal fire-tube boilers need to be cleaned more regularly than vertical fire-tube boilers because of greater accumulation of soot and fly ash.

Wood Stoves and Flue

When looking at wood stoves for the first time, some people use their experience of open fires as a starting point, but there are many differences between them. There is also a significant difference between the burning of solid mineral fuels (e.g. coal) and the burning of wood. As fuels, both coal and smokeless fuel require a considerable air supply from beneath and above the grate to burn the volatile gases given off from the firebed. Wood, however, requires very little air from beneath the grate once combustion is established but needs sufficient air in the gas zone above the fire or in a secondary chamber to ensure efficient combustion. An open fire is at best 20% efficient (37% with smokeless fuel) as large volumes of warm air from the room are lost up the chimney along with the smoke from the fire. Modern 'clean burn' wood stoves, by comparison, may be around 80% efficient. These stoves can be used to heat single rooms or small houses and are available with outputs from 3.5 kW to 20 kW.

Modern Wood Stoves

Modern clean burning wood stove.

Wood stove technology has been developed to a point where clean burning and efficient stoves are now commonly available. Operating in an almost sealed enclosure, and with a well-controlled and distributed air supply, a number of these stoves are now approved for use in smoke control areas. Several types of wood stoves are available: those designed to heat a single room (with or without a back boiler) or, when an internal boiler is incorporated, those that provide all of the heating and hot water requirements of a house.

The use of modern designs results in higher combustion temperatures, and produces better fuel economy and fewer solid deposits than earlier designs. This technology provides the almost complete combustion of the tars and creosotes produced, resulting in a self-cleaning viewing window on the stove door and fewer deposits in the flue ways. For best results wood burning should take place on a bed of embers which trap fragments of charred wood and help to ensure the complete combustion of the fuel. Without an ember bed or a flat fire brick base, partially burned fuel would fall through the grate, resulting in reduced fuel efficiency.

Simple Wood Stoves

The basic wood stove produces heat in two ways: by radiation from its hot surfaces and by convection from air drawn in around the stove casing which is discharged from slots on top of the casing. Such stoves can generally heat only one room but in a house designed to be open plan, convection currents can move heat to the different parts of the house. Some inset convector stoves have air distribution outlets from the convection chamber, allowing convected heat to be distributed to other rooms. Most of the heat emitted from this type of wood stove is by radiation. However, 'cool to touch' convection stoves, which heat primarily by convection and not by radiation, are becoming increasingly available.

Wood Stoves with Integral Boilers

Wood stoves incorporating a hot water boiler are available. These are not dissimilar to the traditional back boiler in an open fire. Such a stove can potentially provide all the heating and hot water for a house; however, a significant proportion of the heat is still supplied to the room in which the stove is located, typically in the ratio 45% to the room and 55% to the boiler for a free-standing stove. Inset versions of stoves from some manufacturers have greater proportion of heat output to water because the stove jacket is insulated as part of the installation, and radiation only occurs from the front of the appliance.

Multi-fuel Stoves

Some manufacturer market stoves suitable for burning both coal and wood, but such stoves are usually a compromise design between the different requirements for burning these two fuels. A multi-fuel stove is unlikely to be optimised for either fuel, while stoves optimised for burning one fuel will usually burn other fuels inefficiently. However, some multi-fuel stoves have an additional insert grate for the fossil fuel option to allow efficient operation with both fuels. Problems may arise if both coal and wood are burned together. During combustion coal produces volatile gases containing sulphur while the combustion of even well-seasoned wood produces water vapour. Together, the sulphur and water vapour produce sulphurous acid and sulphuric acid which may corrode cold metal surfaces and shorten the life of the stove. This occurs when water in the flue gases condenses to form a liquid.

Flues

Flue Liners

While brick chimneys are adequate for an open wood fire they should not be used for a wood stove. Closed wood stoves produce a more concentrated smoke and higher moisture content in the flue gases. Condensation can occur when flue gases come into contact with a cold chimney surface, resulting in any residual tars and creosotes condensing in the chimney. These can soak into the brickwork, damaging the chimney, and also run back down to the stove causing a fire hazard. For this reason any existing chimney must be lined.

Several methods of lining a chimney are available, for example inserting a flexible stainless steel flue liner manufactured specifically for solid fuel use, using a pumped refractory concrete lining or adding a clay or concrete liner to the brickwork. The chimney must be thoroughly sealed against gas escape along its whole length, and against rain ingress between flue liner and chimney at the top. Flue pipes should have the same diameter or equivalent cross-sectional area as that of the appliance flue outlet. Flue liners must always be installed within a brickwork chimney. Where a chimney isn't available a convenient method of constructing one is to use prefabricated, insulated, twin wall chimney components. Any work on a chimney is carried out by a registered competent engineer in combination with a stove installation.

Even with a lined and insulated chimney it is important to burn only quality dry timber. The use of damp or unseasoned wood must be avoided and the fire must not be left to slumber for long periods (e.g. overnight). In this situation the stove never reaches its design operating temperature, resulting in incomplete combustion and the production of tars, creosotes and high levels of carbon monoxide (CO) in the flue gases. The tars and creosotes then condense in the liner because there is insufficient heat in the flue gases to maintain a temperature above their condensation point. This is to be avoided because an insufficient draw from the flue can lead to CO spilling back into the dwelling with serious or even fatal health consequences.

Flue Height

The height of a flue for an appliance rated at less than 50 kW is regulated by the local authority Building Control Department. Building Control Consent or a Building Warrant will be required for all wood stoves in order to ensure compliance with the regulations, and that the flue is correctly matched to the stove. Alternatively, a registered competent engineer who has been approved under a relevant scheme, e.g. HETAS, would be able to self-certify an installation.

A flue which is too short will not produce sufficient negative pressure in the stove to draw the flue gases from the stove, and could allow these gases (including CO) to escape if sufficient combustion air is not available. Building Regulations require and specify minimum flue height which should be sufficient. A flue which is too tall will result in excess air flow through the stove which can cause combustion to occur too quickly; this, in turn, can cause the stove to overheat. Excess air flow in the flue will reduce the efficiency of the stove by carrying more heat than necessary up the flue. If this situation arises expert advice should be sought as draught stabilization maybe required.

Combustion Air Supply

An adequate external air supply is essential for the safe operation of the stove. If the air supply in the room in which the stove is installed is insufficient the stove's flue could produce enough suction to place the room under significant negative pressure. Once a balance pressure has been reached the stove will be partially deprived of oxygen, resulting in the production of carbon monoxide (CO) and inefficient combustion. The lack of air will allow products of combustion, including CO, back into the room. Not only will the stove burn fuel inefficiently, increasing fuel consumption, but tar deposits may cause chimney fires and partial blockage of the flue. However, by far the greatest hazard is that the CO released creates an extremely dangerous situation for the occupants and must be avoided.

As well as avoiding negative pressure in the room, the stove and flue should ensure that the combustion chamber in the stove is always under negative pressure to prevent the escape of wood gas. Room volume required for a stove should be a minimum of 4 m^3 per kW of stove rating. For example, in a typical modern living room measuring 5 m by 3 m and with a ceiling height of 2.4 m, a stove rated at no more than 9 kW should be used. In practice, in a well-insulated modern dwelling, a stove of no more than half this size is likely to be required, but in an old solid stone and poorly insulated house a stove of at least this size may be required, and care would be needed to ensure a sufficient air supply to the stove.

A particular issue arises in dwellings with an open plan ground floor, where the kitchen is fitted with extractor fans. In a close to airtight modern building without adequate fresh air inlet for the stove, a kitchen extractor fan can reduce the room pressure to below the recommended –4 pascals. The installer must take these factors into account and then ensure that a large enough permanent opening to the outside air is provided near the stove for combustion air. In all cases the installation must comply with the relevant advice in approved documents and technical standards.

System Design

All batch fed wood boilers and most wood stoves with integral boilers should have an external heat store to take the heat from the boiler when a load is not present. If this store is large enough it can provide for the daily peak loads experienced when a heating system is started each day, and also provide hot water for sinks and showers.

Various rules of thumb exist for sizing thermal stores associated with stoves and other batch fed boilers. Following formula to calculate the size of storage tank required for a batch fed boiler within a stove:

$$V_{st} = 15\, T_C\, Q_N \left(1 - 0.3\frac{Q_H}{Q_{min}}\right)$$

where, V_{st} is the volume of the tank, T_C is the combustion time at rated heat output, Q_N is the nominal heat output, Q_H is the building heat load and Q_{min} is the minimum output of the boiler.

In practice, however, the use of this equation may be unnecessary or too onerous. A commonly encountered rule of thumb suggests that the storage volume should be no less than 50 litres per kW (l/kW) of boiler rating. A range of between 50 l/kW and 75 l/kW is also often quoted although the upper end of this range may be too large for many applications and would, in turn, limit the size of boiler which could be fitted in a batch fed system. As with the design of all wood burning systems it is important not to over-size the boiler for reasons of capital cost and system efficiency.

Stove with Boiler

Heating System Design

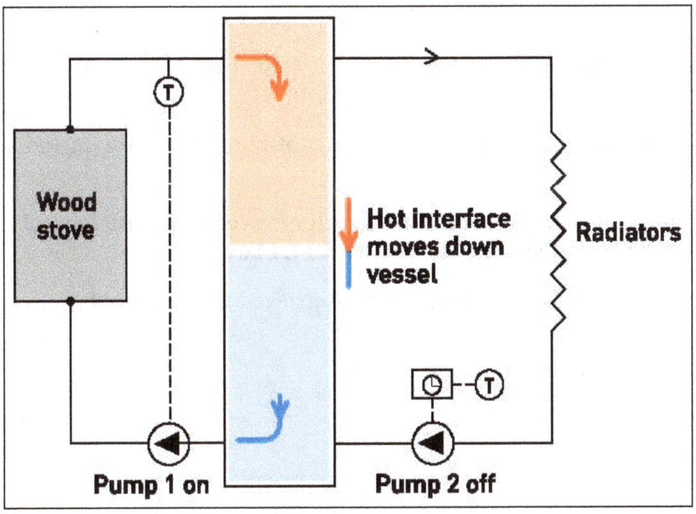

Wood stove burning: No load store charges.

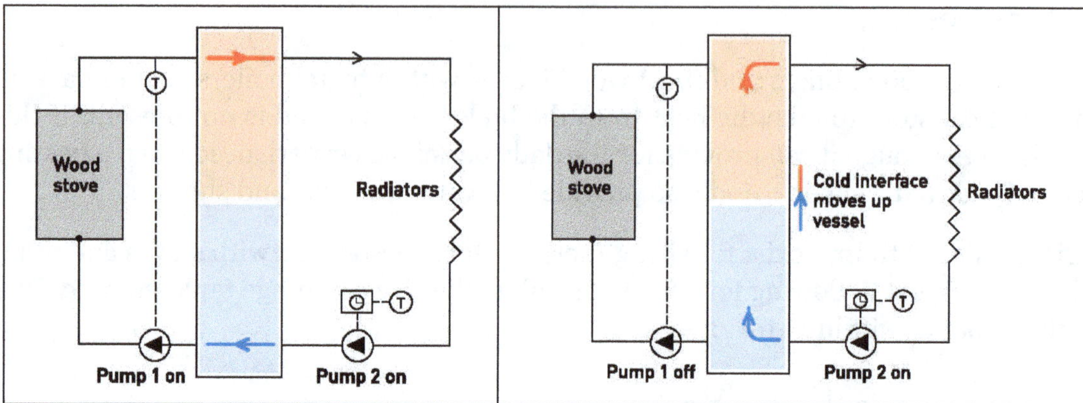

Wood stove on: Load supplied from stove.	Wood stove off: Load supplied from store discharges.

Figures show the essential elements of a heating system based on a wood boiler. The boiler should be allowed to heat up once the fire has been laid and before water is circulated around the system. This prevents condensation forming on the outside of the boiler. The simplest way to achieve the necessary level of control is to position a thermostat on the flow pipe from the boiler to control the boiler pump (pump 1). As the stove heats up, the water in the boiler heats up and convective circulation starts from the boiler to the thermal store. When the temperature of the water leaving the boiler is sufficient the thermostat switches on pump 1, dumping the heat produced into the thermal store which then charges up. If no load is present the heat will be stored. Once a demand for heat occurs the heating pump (pump 2) switches on (shown here in figure under the control of a room thermostat and time switch) and draws heat from the thermal store. If the stove is producing heat at this time, the thermal store is effectively bypassed as shown. However, if the stove is not operating the thermal store meets the heat load for as long as hot water is available in the store, as shown in figure.

References

- Thermal-Biomass-Conversion-A-Review-341017549: researchgate.net, Retrieved 13, February 2020
- Biochemical-conversion-of-biomass-and-status-of-biomass-energy-in-india: ijert.org, Retrieved 05, August 2020
- Biochemical-Conversion-Process-of-Producing-Bioethanol-from-Lignocellulosic-Biomass-265806097: researchgate.net, Retrieved 27, March 2020
- Biomass-Boilers-Technology-Guide: seai.ie, Retrieved 19, June 2020

Chapter 3

Biofuel and its Types

The fuel which is derived through various processes from biomass is termed as biofuel. Some of the prominent types of biofuels are bioethanol, biodiesel, green diesel and aviation biofuel. This chapter closely examines the key aspects of these types of biofuels as well as hydrogen production using microbial biomass.

Biofuel refers to liquid or gaseous fuels that are predominantly produced from biomass. A variety of fuels can be produced from biomass resources including liquid fuels, such as ethanol, methanol, biodiesel, Fischer-Tropsch diesel, and gaseous fuels, such as hydrogen and methane.

Resource Base for Biofuels Production

The biomass resource base for biofuels production is composed of a wide variety of forestry and agricultural resources, industrial processing residues, municipal solid wastes and urban wood residues.

The agricultural resources include grains used for biofuels production, animal manures and residues, and crop residues derived primarily from corn and small grains (e.g., wheat straw). A variety of regionally significant crops, such as cotton, sugarcane, rice, and fruit and nut orchards can also be a source of crop residues.

The forest resources include residues produced during the harvesting of forest products, fuelwood extracted from forestlands, residues generated at primary forest product processing mills, and forest resources that could become available through initiatives to reduce fire hazards and improve forest health. Municipal and urban wood residues are widely available and include a variety of materials yard and tree trimmings, land-clearing wood residues, wooden pallets and organic wastes.

Globally, biofuels are commonly used to power vehicles, heat homes, and for cooking. Biofuel industries are expanding in Europe, Asia and the Americas. Biofuels are generally considered as offering many priorities, including sustainability, reduction of greenhouse gas emissions, regional development, social structure and agriculture, and security of supply.

First-generation Biofuels

First-generation biofuels are made from sugar, starch, vegetable oil, or animal fats using conventional technology. The basic feedstocks for the production of first-generation biofuels come from agriculture and food processing. The most common first-generation biofuels are:

- Biodiesel: Extraction with or without esterification of vegetable oils from seeds of plants like soybean, oil palm, oilseed rape and sunflower or residues including animal fats derived from rendering applied as fuel in diesel engines.
- Bioethanol: Fermentation of simple sugars from sugar crops like sugarcane or from starch crops like maize and wheat applied as fuel in petrol engines.
- Bio-oil: Thermo-chemical conversion of biomass. A process still in the development phase.
- Biogas: Anaerobic fermentation or organic waste, animal manure, crop residues an energy crops applied as fuel in engines suitable for compressed natural gas.

First-generation biofuels can be used in low-percentage blends with conventional fuels in most vehicles and can be distributed through existing infrastructure. Some diesel vehicles can run on 100 % biodiesel, and 'flex-fuel' vehicles are already available in many countries around the world.

Second-generation Biofuels

Second-generation biofuels are derived from non-food feedstock including lignocellulosic biomass like crop residues or wood. Two transformative technologies are under development.

- Biochemical: Modification of the bioethanol fermentation process including a pre-treatment procedure
- Thermochemical: Modification of the bio-oil process to produce syngas and methanol, Fisher-Tropsch diesel or dimethyl ether (DME).

Advanced conversion technologies are needed for a second generation of biofuels. The second generation technologies use a wider range of biomass resources agriculture, forestry and waste materials. One of the most promising second-generation biofuel technologies lignocellulosic processing (e. g. from forest materials) is already well advanced. Demonstration plants have already been established in Denmark, Spain and Sweden.

Etihad Airways flight from Seattle to Abu Dhabi in January 2012 was the first in the Middle East to be powered by sustainable biofuel.

Third-generation Biofuels

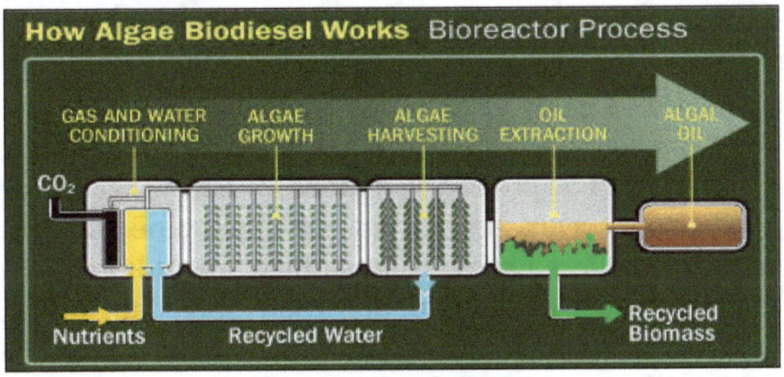

Bioreactor Process.

Third-generation biofuels may include production of bio-based hydrogen for use in fuel cell vehicles from microalgae. The production of algae biofuel, also called Oilgae, is supposed to be low cost and high-yielding – giving up to nearly 30 times the energy per unit area as can be realized from current, conventional 'first-generation' biofuel feedstocks. Algaculture can be an attractive route to making vegetable oil, biodiesel, bioethanol and other biofuels.

Biothanol

The principle fuel used as a petrol substitute for road transport vehicles is bioethanol. Bioethanol fuel is mainly produced by the sugar fermentation process, although it can also be manufactured by the chemical process of reacting ethylene with steam. The main sources of sugar required to produce ethanol come from fuel or energy crops. These crops are grown specifically for energy use and include corn, maize and wheat crops, waste straw, willow and popular trees, sawdust, reed canary grass, cord grasses, jerusalem artichoke, myscanthus and sorghum plants. There is also ongoing research and development into the use of municipal solid wastes to produce ethanol fuel.

Ethanol or ethyl alcohol (C_2H_5OH) is a clear colourless liquid, it is biodegradable, low in toxicity and causes little environmental pollution if spilt. Ethanol burns to produce carbon dioxide and water. Ethanol is a high octane fuel and has replaced lead as an octane enhancer in petrol. By blending ethanol with gasoline we can also oxygenate the fuel mixture so it burns more completely and reduces polluting emissions. Ethanol fuel blends are widely sold in the United States. The most common blend is 10% ethanol and 90% petrol (E10). Vehicle engines require no modifications to run on E10 and vehicle warranties are unaffected also. Only flexible fuel vehicles can run on up to 85% ethanol and 15% petrol blends (E85).

Benefits of Bioethanol

Bioethanol has a number of advantages over conventional fuels. It comes from a renewable resource i.e. crops and not from a finite resource and the crops it derives from can grow well in the UK (like cereals, sugar beet and maize). Another benefit over fossil fuels is the greenhouse gas emissions. The road transport network accounts for 22% of all greenhouse gas emissions and through the use of bioethanol, some of these emissions will be reduced as the fuel crops absorb the CO_2 they emit through growing. Also, blending bioethanol with petrol will help extend the life of the diminishing oil supplies and ensure greater fuel security, avoiding heavy reliance on oil producing nations. By encouraging bioethanol's use, the rural economy would also receive a boost from growing the necessary crops. Bioethanol is also biodegradable and far less toxic that fossil fuels. In addition, by using bioethanol in older engines can help reduce the amount of carbon monoxide produced by the vehicle thus improving air quality. Another advantage of bioethanol is the ease with which it can be easily integrated into the existing road transport fuel system. In quantities up to 5%, bioethanol can be blended with conventional fuel without the need of engine modifications. Bioethanol is produced using familiar methods, such as fermentation, and it can be distributed using the same petrol forecourts and transportation systems as before.

Bioethanol Production

The process of making corn into ethanol is a multistep process. The first step is to milling the corn. It can be done by dry milling or wet milling. Figures show the process steps for each wet and dry milling. For wet milling, the corn kernels are broken down into starch, fiber, corn germ, and protein by heating in the sulfurous acid solution for 2 days. The starch is separated and can produce ethanol, corn syrup, or food-grade starch.

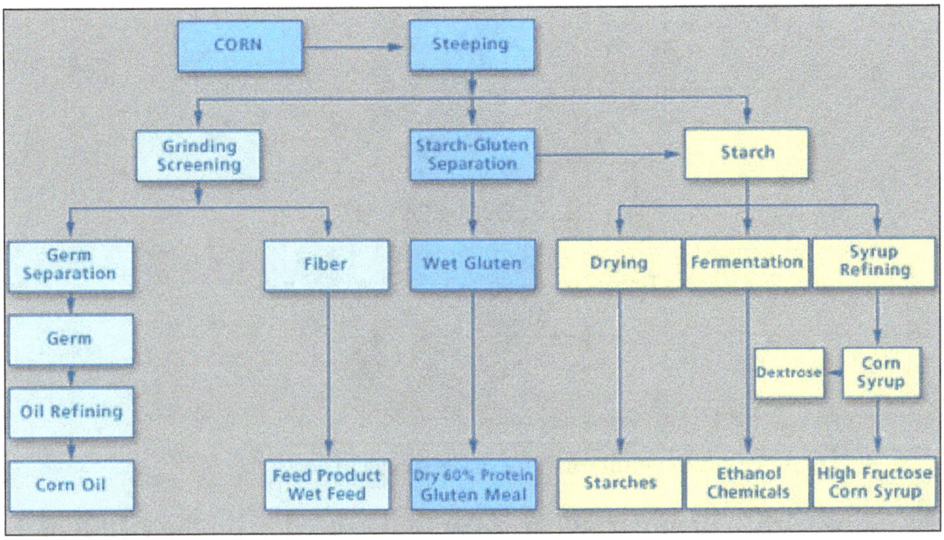

Wet Milling Process.

The wet milling process also produces additional products including feed, corn oil, gluten meal, and gluten feed. Dry milling is a simpler process than wet milling, but it also produces fewer products. The main products of dry milling are ethanol, CO_2, and dried distiller grain with solubles (DDGS). The five steps are grinding, cooking and liquefaction, saccharification, fermentation and distillation.

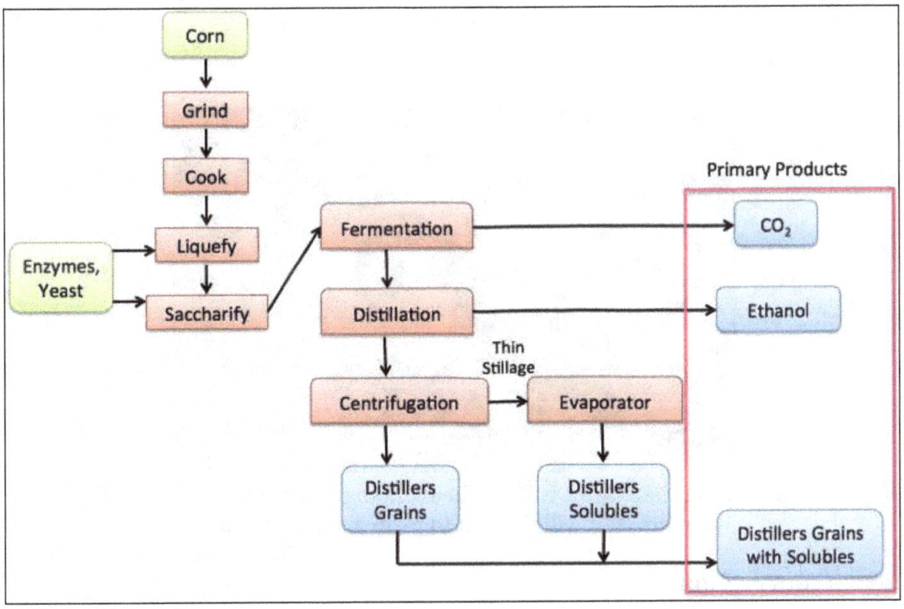

Dry grind ethanol process.

Grinding

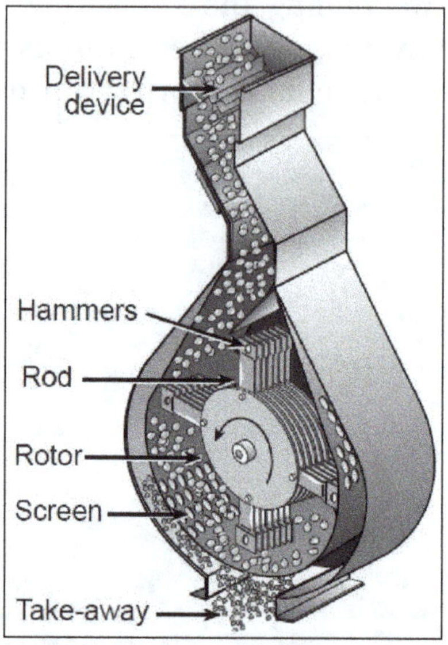

Hammermill for dry grinding of corn.

For dry grinding corn, a hammermill or roller mill is used to do the grinding. Figure is a schematic of a hammermill with corn being put through it. The hammers are attached to rods that turn on a rotor. As the rotor turns, the feed (corn in this case) is hammered against the wall. A screen at the bottom allows particles that are small enough to leave the unit and keep in the larger particles to continue to be hammered until all the material is in the correct size range. The grinding helps to break the tough outer coatings of the corn kernel, which will increase the surface area of the starch. Once the corn is broken down, it is mixed/slurried with heated water to form a mash or slurry.

Cooking and Liquefaction

Corn starch mixed with water is being poured into a sauce mixture; as it heats, it will thicken to form a sauce or gravy.

Once the corn slurry (mash) is made, it goes through cooking and liquefaction. The cooking stage is also called gelatinization. Water interacts with the starch granules in the corn when the temperature is >60 °C and forms a viscous suspension. Have you ever cooked with cornstarch to make thick gravy? Figure shows a picture of starch mixed with water being poured into a heated sauce as it cooks. It will thicken with heat.

The liquefaction step is actually partial hydrolysis that lowers the viscosity. It is essentially breaking up the longer starch chains into smaller chains. One way to measure this is to look at dextrose equivalents (DE), or a measure of the amount of reducing sugars present in a sugar product, relative to glucose, expressed as a percentage on a dry basis. Dextrose is also known as glucose, and dextrose equivalent is the number of bonds cleaved compared to the original number of bonds. The equation is:

$$\text{Equation: } 100 \times \frac{\text{number of bonds cleaved}}{\text{number of original bonds}}$$

Pure glucose (dextrose): DE = 100

Maltose: DE = 50

Starch: DE = 0

Dextrins: DE = 1 through 13

Dextrins are a group of low molecular weight carbohydrates produced by hydrolysis of starch or glycogen. Dextrins are mixtures of polymers of D-glucose units linked by α (1,4) or α (1,6) glycosidic bonds. Dextrins are used in glues and can be a crispness enhancer for food processing.

Maltodextrin: DE = 3 through 20

Maltodextrin is added to beer.

Starch hydrolysis is where water reacts with the sugar to break the sugar down and form glucose. The water breaks into the H^+ and OH^- ions to interact with the starch as it breaks down.

In order to accomplish liquefaction, the reaction must take place under certain conditions. The pH of the mash is maintained in the range of 5.9-6.2, and ammonia and sulfuric acid are added to the tank to maintain the pH. About one-third of the required type of enzyme, α-amylase, can be added to the mash before jet cooking (2-7 minutes at 105-120 °C) to improve the flowability of the mash. The jet cooking serves as a sterilization step to avoid bacterial contamination during the fermentation step later on. At this stage, shorter dextrins are produced but are not yet glucose.

Three types of processes can be utilized for liquefaction. Figure shows the three options. Process 1 is where the α-amylase is added and the material is incubated at 85-95 °C. Process 2 has the mash in the jet cooker at 105-120 °C for 2-7 minutes, then flows to a flash tank at 90 °C. α-Amylase is added three hours later. The third option, Process 3, adds the α-amylase, the heats in the jet cooker at 150 °C, followed by flow to the flash tank at 90 °C and adding more α-amylase.

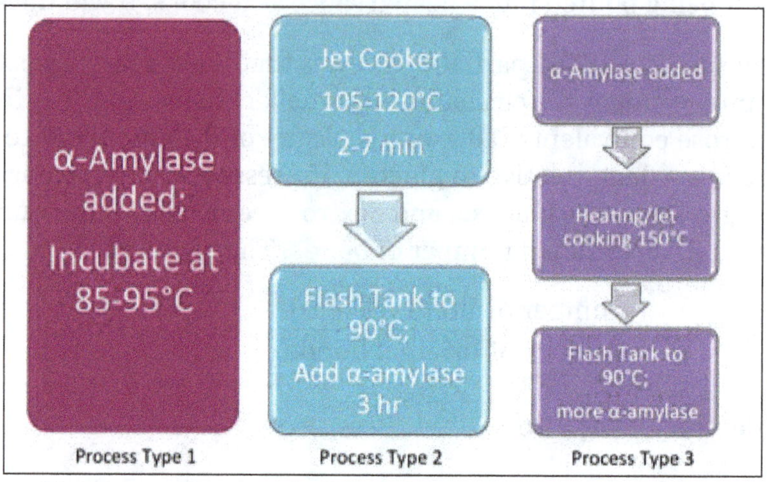

The three option types for liquefaction processing of corn mash.

The α-amylase for liquefaction acts on the internal α (1,4) glycosidic bonds to yield dextrins and maltose (glucose dimers). A type of α-amylase exists in the saliva of humans; a different α-amylase is utilized by the pancreas. Figure below shows one type of α-amylase. The α-amylase works a little faster than the β-amylase, and the β-amylase works on the second α (1,4) glycosidic bond so that maltose is formed. β-amylase is part of the ripening process of fruit increasing the sweetness of fruit as it ripens.

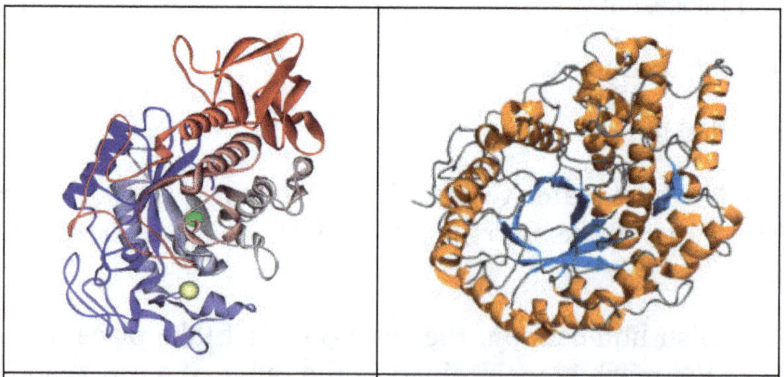

Schematic of an α-amylase. | Schematic of a β-amylase.

Saccharification

The next step in the process of making ethanol is saccharification. Saccharification is the process of further hydrolysis to glucose monomers. A different enzyme is used, called a glucoamylase (also known by the longer name amyloglucosidase). It cleaves both the α (1,4) and α (1,6) glycosidic bonds from dextrin ends to form glucose. The optimum conditions are different from the previous step and are at a pH of 4.5 and a temperature of 55-65 °C. Figure shows a schematic of the glucoamylase, which is also called a ϒ-amylase. There are a wide variety of amylase enzymes available that are derived from bacteria and fungi. Table shows different enzymes, their source, and the action of each.

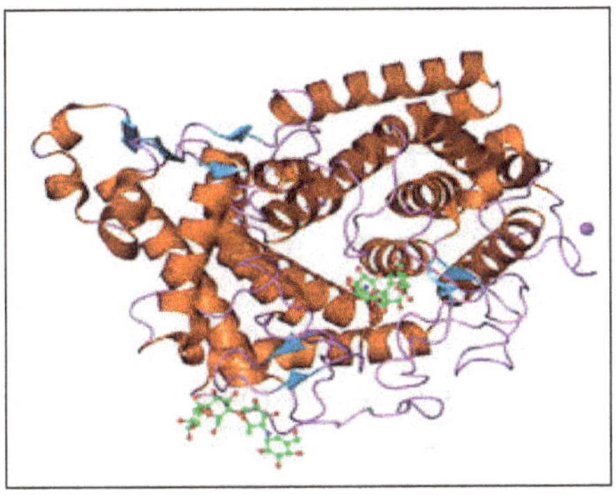

Schematic of a glucoamylase (aka a ϒ-amylase).

Table: Different enzymes used in starch depolymerization.

Enzyme	Source	Action
α-Amylase	Bacillus amyloliquefaciens	Only α-1,4-oligosaccharide links are cleaved to give a-dextrins and predominantly maltose (G2), G3, G6, and G7 oligosaccharides.
	B. licheniformis	Only α-1,4-oligosaccharide links are cleaved to give a-dextrins and predominantly maltose, G3, G4, and G5 oligosaccharides.
	Aspergillus oryzae, A. niger	Only α-1,4 oligosaccharide links are cleaved to give a-dextrins and predominantly maltose and G3 oligosaccharides.
Saccharifying a-amylase	B. subtilis (amylosaccharitícus)	Only α-1,4-oligosaccharide links are cleaved to give a-dextrins with maltose, G3, G4 and up to 50% (w/w) glucose.
β-Amylase	Malted barley	Only α-1,4-links are cleaved, from non-reducing ends, to give limit dextrins and b-maltose.
Glucoamylase	A. niger	α-1,4 and α-1,6-links are cleaved, from the nonreducing ends, to give β-glucose.
Pullulanase	B. acidopullulyticus	Only α-1,6-links are cleaved to give straight-chain maltodextrins.

Some of the newer developed enzymes (granular starch hydrolyzing enzymes – GSHE) allow skipping the liquefaction stage by hydrolyzing starch at low temperatures with cooking.

- Advantages include:
  - Reduced heat/energy.
  - Reduced unit operation (reducing capital and operating costs).
  - Reduced emissions.

    - Higher DDGS. They work by "coring" into starch granules directly without the water swelling/infusion.

- Disadvantages include:
    - Enzymes cost more.
    - Contamination risks.

Fermentation

The final chemical step in producing ethanol from the starch is fermentation. The chemical reaction of fermentation is where 1 mole of glucose yields 2 moles of ethanol and 2 moles of carbon dioxide. The reaction is shown in equation below:

$$C_6H_{12}O_6 \rightarrow 2\,C_2H_6OH + 2\,CO_2$$

To cause fermentation to take place, yeast is added. Common yeast to use is saccharomyces cerevisiae, which is a unicellular fungus. The reaction takes place at 30-32 °C for 2-3 days in a batch process. Supplemental nitrogen is added as ammonium sulfate ($(NH_4)_2SO_4$) or urea. A protease can be used to convert proteins to amino acids to add as an additional yeast nutrient. Virginiamycin and penicillin are often used to prevent bacterial contamination. The carbon dioxide produced also lowers pH, which can reduce the contamination risk. Close to 90-95% of the glucose is converted to ethanol.

It is possible to do saccharification and fermentation in one step. It is called Simultaneous Saccharification and Fermentation (SSF), and both glucoamylase and yeast are added together. It is done at a lower temperature than saccharification (32-35 °C), which slows the hydrolysis into glucose. As glucose is formed, it is fermented, which reduces enzyme product inhibition. It lowers initial glucose concentrations, lowers contamination risk, lowers energy requirements, and produces higher yields of ethanol. Because SSF is done in one unit, it can improve capital costs and save residence time.

Distillation and Increase of Ethanol Concentration

The last phase of ethanol production is the processing of ethanol to increase the ethanol concentration. Downstream from the fermenters, the ethanol concentration is 12-15% ethanol in water. Crude oil must be distilled into various boiling fractions to separate the oil into useable products. Distillation is a process to separate components using heat and specially designed towers to keep the liquid flowing downward and the vapors being generated to flow upwards. Water boils at 100 °C, while ethanol boils at 78 °C. However, because water and ethanol evaporate at a lower temperature than their boiling points, and because they both have OH functional groups that are attracted to each other, ethanol and water molecules are strongly bound to each other and form an azeotrope together. That just means that you cannot completely separate ethanol from water – the ethanol fraction will contain about 5% water and 95% ethanol when you get to the end of the distillation process. Figure shows a schematic of a distillation unit. You don't want water in gasoline as you drive, because it prevents efficient combustion.

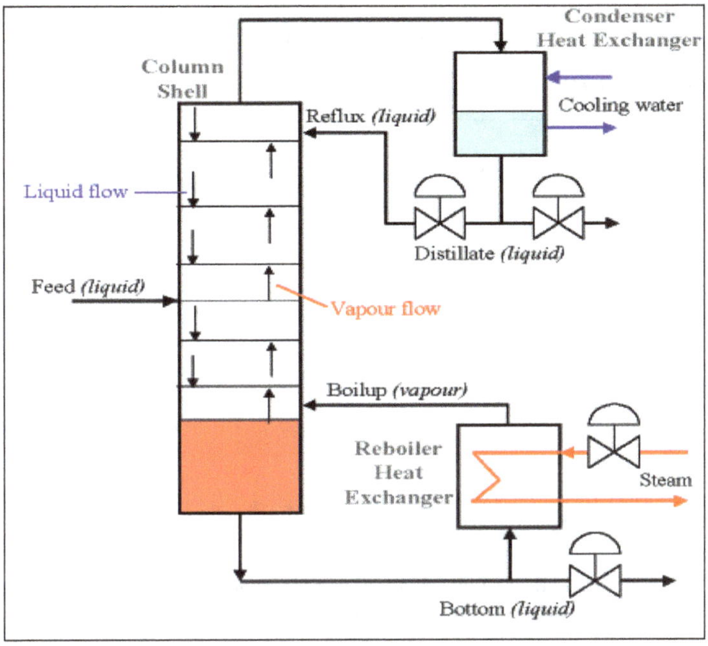

Distillation unit for increasing concentration of ethanol.

Do you want water in your ethanol if you use it as a fuel? The answer is no, so you must use an additional method to remove all the water from ethanol. The method is called dehydration. The unit that is used is called a molecular sieve, and the material used in it is called zeolite. Under these conditions, the zeolite absorbs the water into it, but the ethanol will not go into the zeolite. They use what is called a pressure-swing adsorption unit. The unit is designed to run in two modes. At high pressure, the ethanol is dehydrated in Unit 1, and at low pressure, anhydrous ethanol is fed through to remove the water from Unit 2.

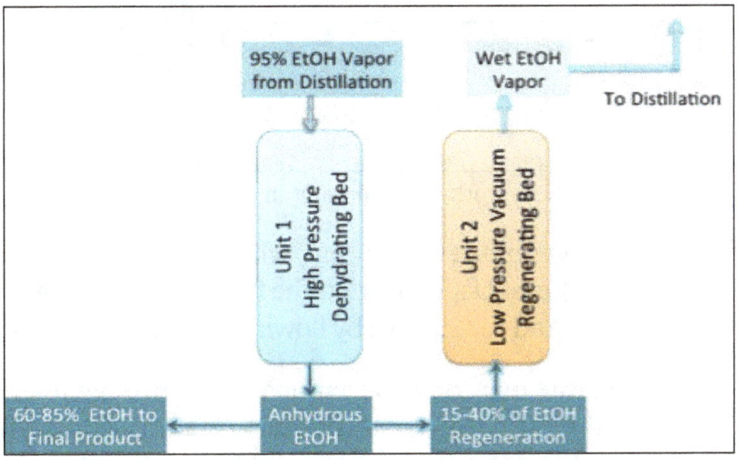

The first unit is the dehydrator to remove water while the second unit is having the water removed.

When the zeolite sieve has absorbed all the water, Unit 1 is switched to become the low pressure regenerating bed, and Unit 2 becomes the high-pressure unit. The residence time for the process is 3-10 minutes. The zeolite for this process is a highly ordered aluminosilicate with well-defined pore sizes that are formed into beads or included in a membrane. The zeolites attract both water and ethanol, but the pore sizes are too small to allow the ethanol to enter.

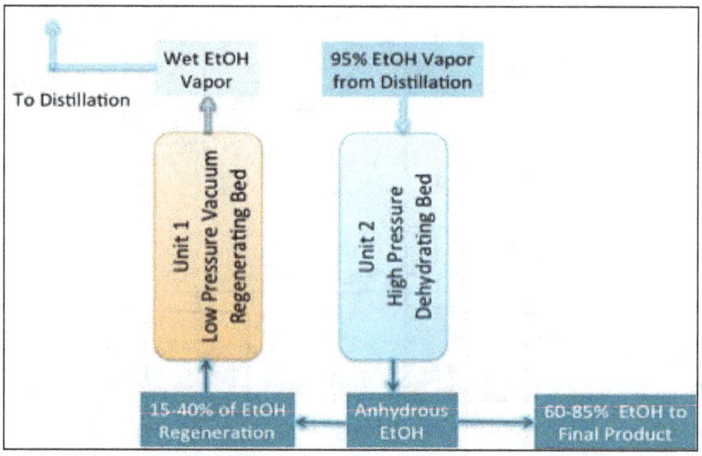

The units switch places because the second bed of zeolite had the moisture removed and now acts as a dehydrator.

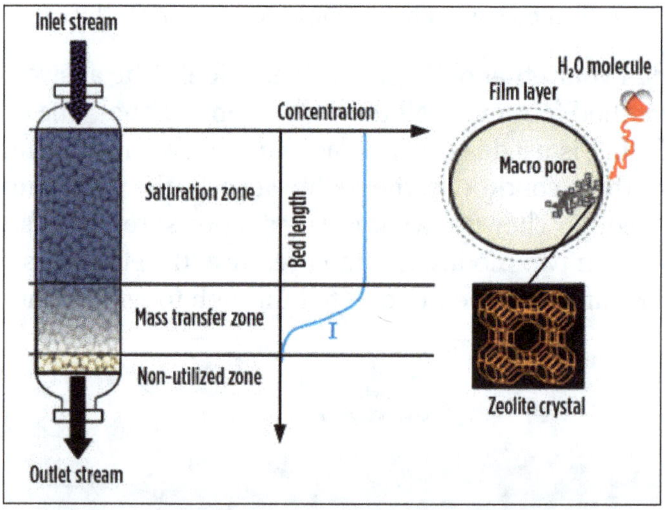

How the sieve works to keep water in and ethanol out.

The pore size of the zeolite membrane is 0.30 nm, while the size of the water molecule is 0.28 nm and the ethanol 0.44 nm. Depending on the type of unit, the membrane or beads can be regenerated using heat and vacuum, or by flowing the pure ethanol through the unit.

So once we have fermented the material to ethanol, it goes through a series of processes to obtain the products in the form that we want them. Figure is a schematic of product recovery, and figure shows the definitions of some of the terminology.

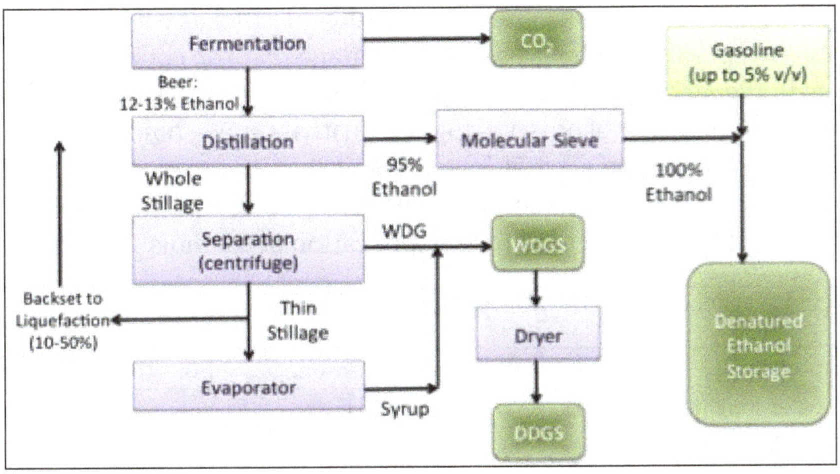

Product recovery diagram of ethanol and other products.

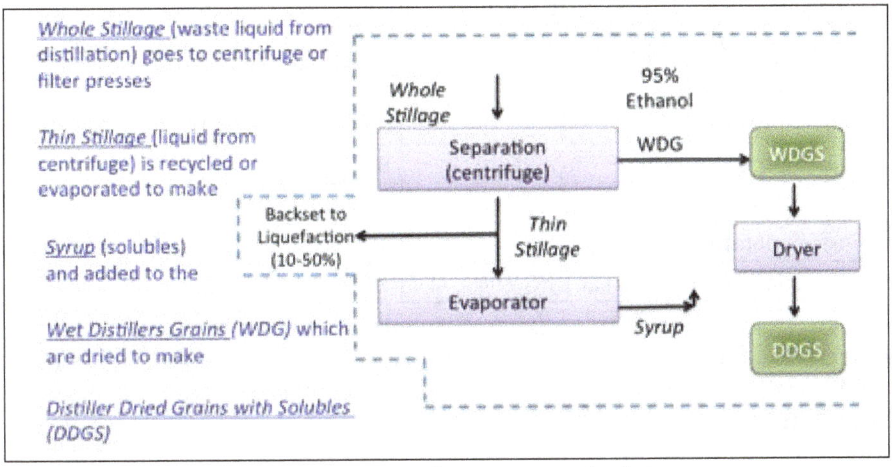

Product separation/recovery terminology.

Corn has 62% starch, 19% protein, 4% oil, and 15% water. If you look at the products on a dry basis (you don't look at the water like a product), 73% of the corn is starch and 27% is protein, fiber, and oil. For every bushel of corn, realistically you'll generate 2.8 gallons of ethanol, ~17 lbs of CO_2, and ~17 lbs of DDGS.

Properties of Ethanol

Bioethanol has much lower energy content than gasoline (about two-third of the energy content of the latter on a volume base). This means that, for mobility applications, for a given tank volume, the range of the vehicle is reduced in the same proportion. The octane number of ethanol is higher than that for petrol; hence ethanol has better anti-knock characteristics. This better quality of the fuel can be exploited if the compression ratio of the engine is adjusted accordingly. This increases the fuel efficiency of the engine. The oxygen content of ethanol also leads to a higher efficiency, which results in a cleaner combustion process at relatively low temperatures.

The Reid vapour pressure, a measure for the volatility of a fuel, is very low for ethanol. This indicates a slow evaporation, which has the advantage that the concentration of evaporative emissions in the air remains relatively low. This reduces the risk of explosions. However, the low vapour pressure of ethanol, together with its single boiling point, is disadvantageous with regard to engine start at low ambient temperatures. Without aids, engines using ethanol cannot be started at temperatures below 20 °C. Cold start difficulties are the most important problem with regard to the application of alcohols as automotive fuels.

Table: Fuel properties of gasoline, bioethanol and ETBE.

Fuel properties	Gasoline	Bioethanol	ETBE
Molecular weight [kg/kmol]	111	46	102
Density [kg/l] at 15°C	0.75	0.80-0.82	0.74
Oxygen content [wt-%]		34.8	
Lower Calorific Value [MJ/kg] at 15°C	41.3	26.4	36
Lower Calorific Value [MJ/l] at 15°C	31	21.2	26.7
Octane number (RON)	97	109	118
Octane number (MON)	86	92	105
Cetane number	8	11	-
Stoichiometric air/fuel ratio [kg air/kg fuel]	14.7	9.0	-
Boiling temperature [°C]	30-190	78	72
Reid Vapour Pressure [kPa] at 15°C	75	16.5	28

Applications of Ethanol

Ethanol can be used:

- As a transport fuel to replace gasoline.
- As a fuel for power generation by thermal combustion.
- As a fuel for fuel cells by thermochemical reaction.
- As a fuel in cogeneration systems.
- As a feedstock in the chemicals industry.

Ethanol is best used in spark-ignition engines because of its high octane rating. Due to its poor ignition quality (low cetane number), it is less suitable for diesel engines. It is generally impractical to use neat ethanol in spark-ignition engines due to its low vapour pressure and high latent heat of vaporisation which make cold start problematic. The most cost-effective aid is the blending of ethanol with a small proportion of a volatile fuel such as gasoline. Thus, various mixture of bioethanol with gasoline or diesel fuels have been used. The most well-known blends are (by volume):

- E5G to E26G (5-26% ethanol, 95-74% gasoline).

- E85G (85% ethanol, 15% gasoline).
- E15D (15% ethanol, 85% diesel).
- E95D (95% ethanol, 5% water, with ignition improver).

Bioethanol has been extensively tested in light duty flexible fuel vehicles (FFV) as E85G. ETBE is also used in blends of 10-15 % with gasoline to enhance its octane rating and reduce emissions. Blends of gasoline with up to 22% ethanol (E22G) can be used in spark ignition engines without any material or operating problems. Blends of diesel with up to 15% ethanol (E22D) do not introduce any technical engine problem and require no ignition improver.

Table: Bioethanol production costs in the EU-25 + Bulgaria, Romania.

	Wheat based			Beet based		
	€/L	€/GJ	€/toe	€/L	€/GJ	€/toe
Net feedstock cost						
Feedstock	0.40	18.9	790	0.26	12.3	513
Co-product credit	0.15	7.1	296	0.03	1.4	59
Subtotal feedstock cost	0.25	11.8	493	0.23	10.9	454
Conversion costs	0.28	13.3	553	0.22	10.4	434
Blending costs (incl. adaptation of gasoline)	0.05	2.4	99	0.05	2.4	99
Distribution costs	0.01	0.5	20	0.1	4.7	197
Total costs at petrol station	0.59	27.9	1165	0.6	28.4	1184

Biodiesel

Biodiesel is an alternative fuel similar to conventional or 'fossil' diesel. Biodiesel can be produced from straight vegetable oil, animal oil/fats, tallow and waste cooking oil. The process used to convert these oils to Biodiesel is called transesterification. The largest possible source of suitable oil comes from oil crops such as rapeseed, palm or soybean. Rapeseed represents the greatest potential for biodiesel production.

Most biodiesel produced at present is produced from waste vegetable oil sourced from restaurants, chip shops, industrial food producers such as Birdseye etc. Though oil straight from the agricultural industry represents the greatest potential source it is not being produced commercially simply because the raw oil is too expensive. After the cost of converting it to biodiesel has been added on it is simply too expensive to compete with fossil diesel. Waste vegetable oil can often be sourced for free or sourced already treated for a small price. (The waste oil must be treated before conversion to biodiesel to remove impurities). The result is Biodiesel produced from waste vegetable oil can compete with fossil diesel.

Rapeseed fields produce vivid colours.

Benefits of Biodiesel

Biodiesel has many environmentally beneficial properties. The main benefit of biodiesel is that it can be described as 'carbon neutral'. This means that the fuel produces no net output of carbon in the form of carbon dioxide (CO_2). This effect occurs because when the oil crop grows it absorbs the same amount of CO_2 as is released when the fuel is combusted. In fact this is not completely accurate as CO_2 is released during the production of the fertilizer required to fertilize the fields in which the oil crops are grown. Fertilizer production is not the only source of pollution associated with the production of biodiesel, other sources include the esterification process, the solvent extraction of the oil, refining, drying and transporting. All these processes require an energy input either in the form of electricity or from a fuel, both of which will generally result in the release of greenhouse gases. To properly assess the impact of all these sources requires use of a technique called life cycle analysis. Biodiesel is rapidly biodegradable and completely non-toxic, meaning spillages represent far less of a risk than fossil diesel spillages. Biodiesel has a higher flash point than fossil diesel and so is safer in the event of a crash.

Biodiesel Production

Biodiesel can be produced from straight vegetable oil, animal oil/fats, tallow and waste oils. There are three basic routes to biodiesel production from oils and fats:

- Base catalyzed transesterification of the oil.
- Direct acid catalyzed transesterification of the oil.
- Conversion of the oil to its fatty acids and then to biodiesel.

Almost all biodiesel is produced using base catalyzed transesterification as it is the most economical process requiring only low temperatures and pressures and producing a 98% conversion yield. The Transesterification process is the reaction of a triglyceride (fat/oil) with an alcohol to form esters and glycerol. A triglyceride has a glycerine molecule as its base with three long chain fatty acids attached. The characteristics of the fat are determined by the nature of the fatty acids attached to the glycerine. The nature of the fatty acids can in turn affect the characteristics of the biodiesel. During the esterification process, the triglyceride is reacted with alcohol in the presence of a catalyst, usually a strong alkaline like sodium hydroxide.

The alcohol reacts with the fatty acids to form the mono-alkyl ester, or biodiesel and crude glycerol. In most production methanol or ethanol is the alcohol used (methanol produces methyl esters, ethanol produces ethyl esters) and is base catalysed by either potassium or sodium hydroxide. Potassium hydroxide has been found to be more suitable for the ethyl ester biodiesel production either base can be used for the methyl ester. A common product of the transesterification process is Rape Methyl Ester (RME) produced from raw rapeseed oil reacted with methanol.

The figure below shows the chemical process for methyl ester biodiesel. The reaction between the fat or oil and the alcohol is a reversible reaction and so the alcohol must be added in excess to drive the reaction towards the right and ensure complete conversion.

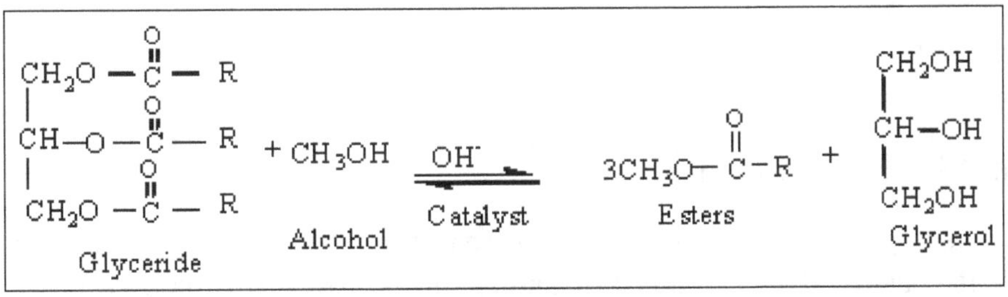

The products of the reaction are the biodiesel itself and glycerol. A successful transesterification reaction is signified by the separation of the ester and glycerol layers after the reaction time. The heavier, co-product, glycerol settles out and may be sold as it is or it may be purified for use in other industries, e.g. the pharmaceutical, cosmetics etc.

Straight vegetable oil (SVO) can be used directly as a fossil diesel substitute however using this fuel can lead to some fairly serious engine problems. Due to its relatively high viscosity SVO leads to poor atomisation of the fuel, incomplete combustion, coking of the fuel injectors, ring carbonisation, and accumulation of fuel in the lubricating oil. The best method for solving these problems is the transesterification of the oil.

The engine combustion benefits of the transesterification of the oil are:

- Lowered viscosity.
- Complete removal of the glycerides.

- Lowered boiling point.
- Lowered flash point.
- Lowered pour point.

Production Process

An example of a simple production flow chart is proved below.

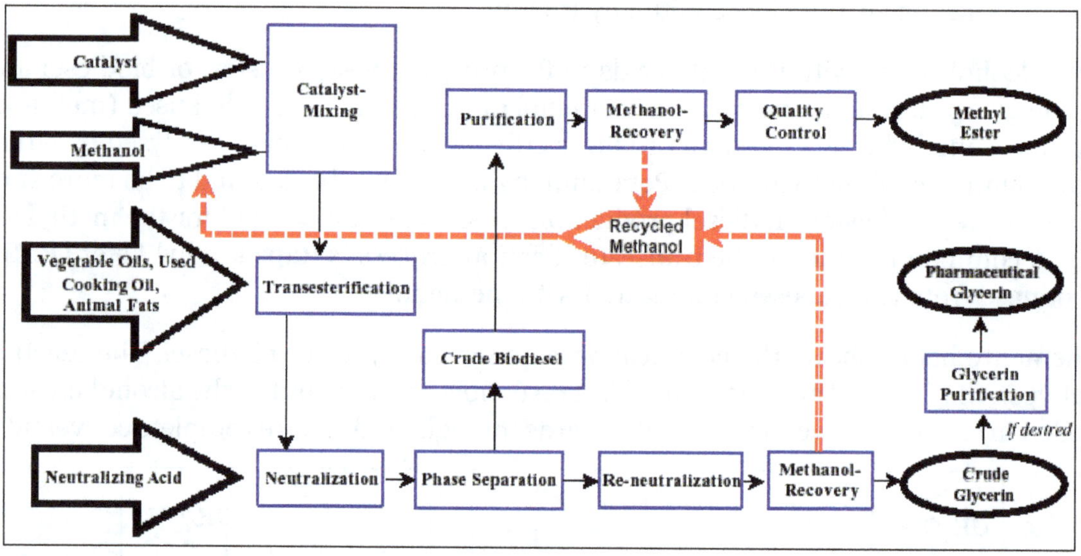

Mixing of Alcohol and Catalyst

The catalyst is typically sodium hydroxide (caustic soda) or potassium hydroxide (potash). It is dissolved in the alcohol using a standard agitator or mixer. The alcohol/catalyst mix is then charged into a closed reaction vessel and the oil or fat is added. The system from here on is totally closed to the atmosphere to prevent the loss of alcohol. The reaction mix is kept just above the boiling point of the alcohol (around 160 °F) to speed up the reaction and the reaction takes place. Recommended reaction time varies from 1 to 8 hours, and some systems recommend the reaction take place at room temperature. Excess alcohol is normally used to ensure total conversion of the fat or oil to its esters. Care must be taken to monitor the amount of water and free fatty acids in the incoming oil or fat. If the free fatty acid level or water level is too high it may cause problems with soap formation and the separation of the glycerin by-product downstream.

Separation

Once the reaction is complete, two major products exist: glycerin and biodiesel. Each has a substantial amount of the excess methanol that was used in the reaction. The reacted mixture is sometimes neutralized at this step if needed. The glycerin phase is much denser than biodiesel phase and the two can be gravity separated with glycerin simply drawn off the bottom of the settling vessel. In some cases, a centrifuge is used to separate the two materials faster.

Alcohol Removal

Once the glycerin and biodiesel phases have been separated, the excess alcohol in each phase is removed with a flash evaporation process or by distillation. In others systems, the alcohol is removed and the mixture neutralized before the glycerin and esters have been separated. In either case, the alcohol is recovered using distillation equipment and is re-used. Care must be taken to ensure no water accumulates in the recovered alcohol stream.

Glycerin Neutralization

The glycerin by-product contains unused catalyst and soaps that are neutralized with an acid and sent to storage as crude glycerin. In some cases the salt formed during this phase is recovered for use as fertilizer. In most cases the salt is left in the glycerin. Water and alcohol are removed to produce 80-88% pure glycerin that is ready to be sold as crude glycerin. In more sophisticated operations, the glycerin is distilled to 99% or higher purity and sold into the cosmetic and pharmaceutical markets.

Methyl Ester Wash

Once separated from the glycerin, the biodiesel is sometimes purified by washing gently with warm water to remove residual catalyst or soaps, dried, and sent to storage. In some processes this step is unnecessary. This is normally the end of the production process resulting in a clear amber-yellow liquid with a viscosity similar to petrodiesel. In some systems the biodiesel is distilled in an additional step to remove small amounts of color bodies to produce a colorless biodiesel.

Product Quality

Prior to use as a commercial fuel, the finished biodiesel must be analyzed using sophisticated analytical equipment to ensure it meets any required specifications. The most important aspects of biodiesel production to ensure trouble free operation in diesel engines are:

- Complete Reaction.
- Removal of Glycerin.
- Removal of Catalyst.

- Removal of Alcohol.
- Absence of Free Fatty Acids.

Life Cycle Assessment

Life Cycle Assessment (or LCA) is a technique to quantitatively assess the environmental impact and the energy requirements of a product or service from its initial raw materials to its final disposal (i.e. cradle to grave). This assessment takes into account any other products or services that may be required to facilitate its use and/or production.

One of the key advantages of using LCA is it allows a direct and fair comparison between two products or services with regards to the environmental and energy impact. In LCA every relevant detail is taken into consideration. This encompasses the distance the raw materials have to be exported to the type of packaging that is used to house the final product.

The level of detail achieved and the quality of the information used determines the accuracy of the assessment. Due to this kind of comparative assessment LCA is an excellent tool to weigh costs and benefits and is therefore a useful aid in decision-making and policy analysis.

When considering LCA for biofuels, it must take into account the evaluation of the energy and global warming costs of producing biodiesel from energy crops and comparing this with other fuels and relevant energy saving measures.

CO_2 Emissions

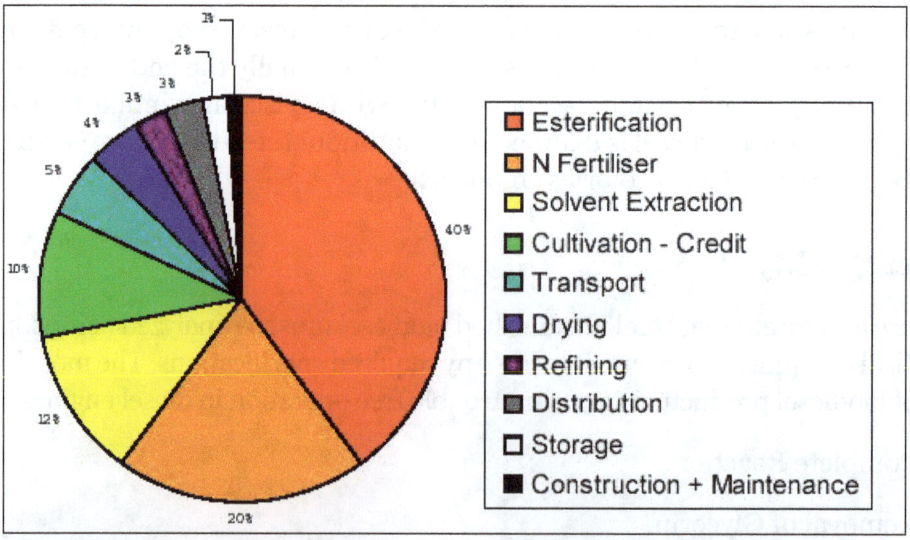

Representative Carbon Dioxide Outputs for Biodiesel by Conventional Production.

Since biodiesel can be described as 'carbon neutral' then any CO_2 emissions associated with it must come from a source outside that of combustion of the fuel. In fact there are many sources of CO_2 associated with biodiesel production. Most of the emissions come from the esterification process, the production of fertilizer and the extraction of the oil from the seed. These CO_2 emissions are not always produced directly from the process but taken from the energy requirement. What this means is that the energy a process uses has an associated emission. For example electricity used in the esterification process may well come from a coal power plant, so the amount of electricity used can be related to the amount of CO_2 released from the plant. It was found that for every ton of biodiesel produced 916 ± 52 kg CO_2 was released into the atmosphere. The chart below shows the proportions of where these emissions come from.

Green House Gas Emissions

GHG emissions can be calculated in just the same way as for CO_2 emissions. In order to group these emissions together each kg of GHG has an equivalent mass of CO_2 based on the 100-year global warming potential. For each ton of biodiesel produced the equivalent of 1,516 ± 88 kg of CO_2 are released.

Energy Requirements

By quantifying the amount of energy required to produce biodiesel it is possible not only to see what processes require the most energy but also to establish an energy balance over the life cycle. I.e. the energy you get out against the energy you put in. For every ton of biodiesel produced 16,269 ± 896 MJ of energy is required. The chart below shows how this is broken down.

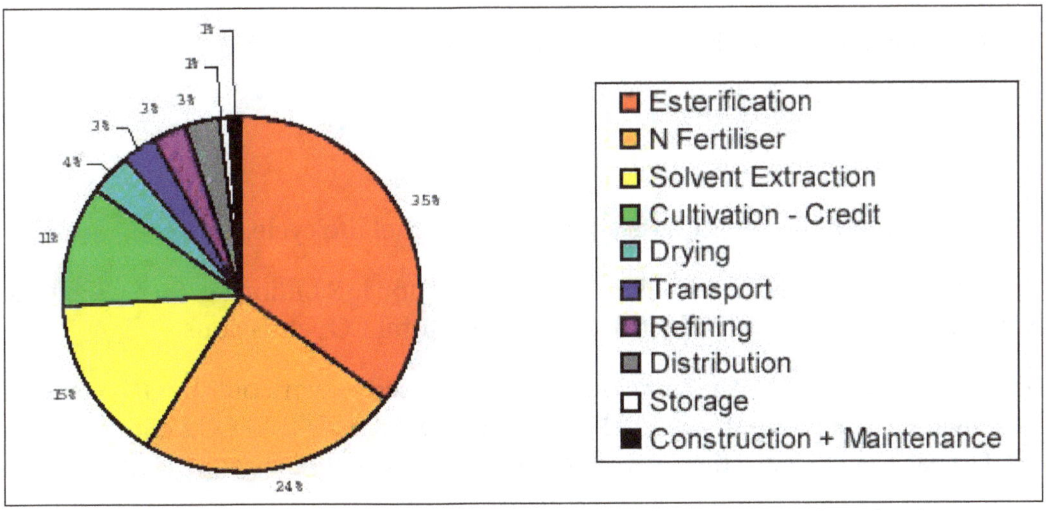

Representative Primary Energy Inputs for Biodiesel by Conventional Production.

Not surprisingly the largest energy demands match up with the CO_2 emissions. A ton of biodiesel will contain around 40,800MJ of energy.

Energy Balance = Energy OUT/Energy IN

= 40,800/16.269

= 2.5

For these number to mean anything however there has to be something against which to compare them. In the case of biodiesel this would be the LCA of fossil diesel.

Biodiesel vs. Fossil Diesel LCA Comparison

1. CO_2 Emissions: For each MJ of biodiesel produced 0.025Kg of CO_2 is released. For each MJ of fossil diesel produced 0.087Kg of CO_2 is released. From this it is possible to work out the emissions savings from the introduction of biodiesel produced. To work out the emission savings the following steps are used:

2% of current diesel consumption = 0.38 billion litres

Energy in 0.38 billion litres of diesel = 0.38 * 45.1 = 17.1 billion MJ

Emissions from 0.38 billion litres of diesel = 17.1 * 0.087

= 1.49billion Kg of CO_2

Biodiesel has lower energy content than fossil diesel so rather than find the emissions for biodiesel from the volume being replaced it is more correct to find the emission from the energy being replaced, therefore:

Emissions from 17.1 billion MJ worth of biodiesel = 17.1 * 0.025

= 0.43 billion Kg of CO_2

Total CO_2 emissions savings = 1.49 – 0.43

= 1.06 billion Kg of CO_2 every year over full fuel life cycle.

In 2000 the UK produced 148 million tonnes of CO_2, or 148 billion kg. So a 2% biodiesel blend would create a 0.72% reduction in the UK's total CO_2 emissions.

2. GHG Emissions: Exactly the same process can be used to work out the GHG emissions and savings. For each MJ of biodiesel produced 0.041Kg of GHG CO2 equivalent is released. For each MJ of fossil diesel produced 0.095Kg of GHG CO2 equivalent is released.

Total GHG saving each year = 0.92 billion Kg of GHG CO_2 equivalent over the full fuel life cycle

3. Energy Requirements: For each MJ of biodiesel produced 0.45 MJ is required. For each MJ of fossil diesel produced 1.26 MJ is required.

Fossil diesel is an energy negative fuel, requiring more energy to produce than it has available within it. Biodiesel is an energy positive fuel containing more energy than was required to produce it.

Green Diesel

Around 50% of the produced crude petroleum in the world is refined into transportation fuels which represent the fastest growing component of the energy sector. This massive use of a non-renewable feedstock leads to a significant footprint on the environment especially in terms of greenhouse gas emissions. In this framework, biomass can be view as the only reliable green energy source that can be converted into liquid transportation fuels in competition with fossil-fuels.

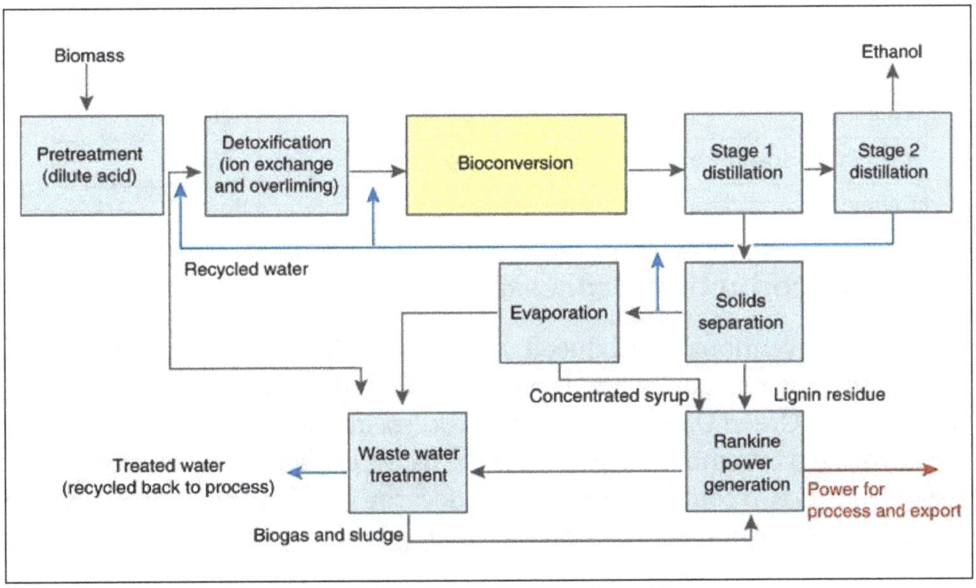

Overview of biofuels production from biomass.

Actually, the most common process to produce biofuels from biomass is the biodiesel production process. As valid and interesting alternative it is possible to produce biofuels for transportation purpose by blending or co-processing of green feedstock within traditional petroleum refineries by properly revamped processes. To this end, researchers has developed a new biofuel production process, called Ecofining, to produce Green Diesel in alternative to fatty acid methyl ester (FAME) also known as biodiesel.

If from one hand, biodiesel has many desirable qualities such as high cetane number, from other hand there are some disadvantages associated with its use, such as poor stability and high solvency, leading to filter plugging problems.

In order to overcome these problems, researchers have developed its new process based on a revamping of a conventional hydroprocessing technology that is widely deployed in refineries and uses the existing refinery infrastructure and fuels distribution system. As already said, this new technology process use widely available vegetable-oil feedstocks to produce a high-cetane, low-gravity, low-aromatics and sulphur-free diesel fuel, known as Green Diesel. Table reports a comparison between physical properties of biodiesel and green diesel.

Table: Comparison of Biodiesel and Green Diesel properties.

	Biodiesel(FAME)	Green Diesel
%O	11	0
Density,g/ml	0.883	0.78
Sulphur Content	<10 ppm	<10 ppm
Heating value (lower), MJ/kg	38	44
% change in NO_x emission	+10	0 to -10
Cloud point C	-5	-5 to -30
Distillation 10-90% pt	340-355	265-320
Cetane	50	80-90

Green Diesel Production Processes

The biodiesel is conventionally produced via transesterification of triglycerides in presence of methanol. The reaction is catalyzed by caustic soda to produce FAME and glycerol as a byproduct. The Green Diesel production process, implementing the hydroprocessing route, uses hydrogen to remove oxygen from the triglyceride molecules. Then, the oxygen is easily removed via two competing reactions: decarboxylation and hydrodeoxygenation. The extent for each reaction depends on the catalyst and process conditions.

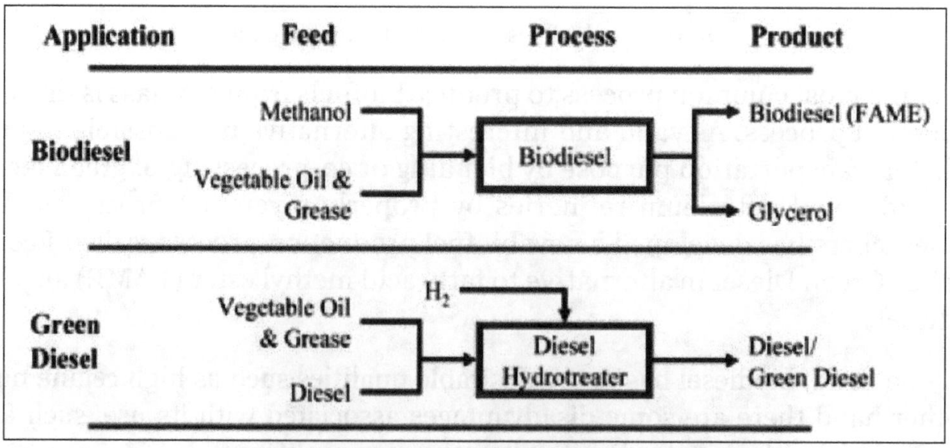

Biodiesel vs. Green Diesel production processes.

The revamping of the hydroprocessing technology to produce green diesel is achievable mainly by two different approaches:

- Co-processing in an existing distillate hydroprocessing unit.
- Building a standalone unit.

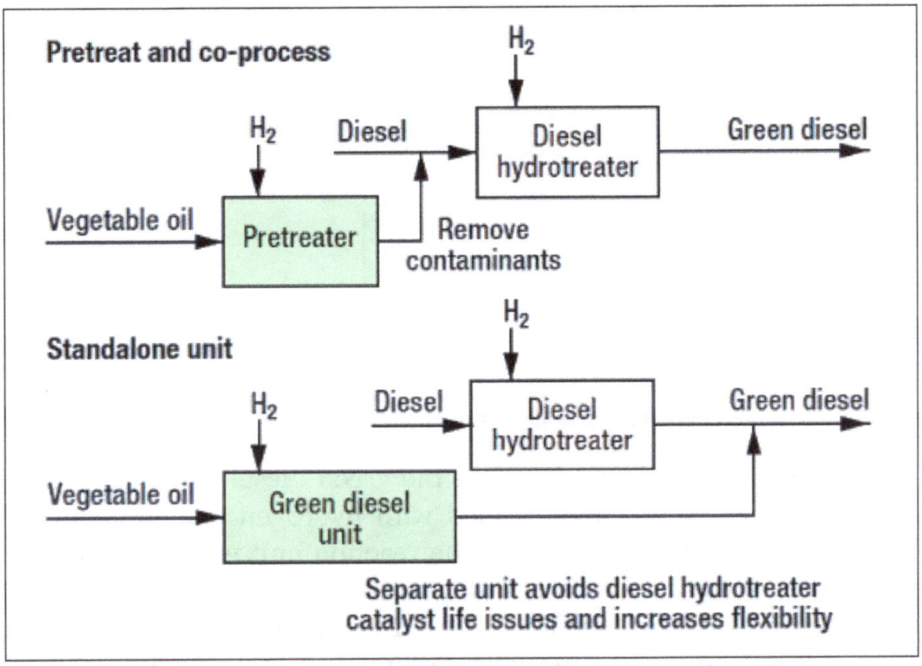

Hydroprocessing scheme for Green Diesel production.

The co-processing route shows good potentially due to the possibility of reuse existing equipments, resulting in a lower implementation cost of the whole plant. Anyway co-processing is affected by some problems related to the processing of vegetable oils which contain trace of metal contaminants such as phosphorous, sodium, potassium and calcium. In fact, in this case a pre-treating reactor may be needed to remove the contaminants by catalytic reactions. Generally existing reactors in the refinery plants not have sufficient catalyst volume to carry out the required reactions. Furthermore, processing vegetable oil reactions are fairly exothermic and may require quench facilities which may not be available. By-products (H_2O, CO, CO_2) coming from the deoxygenation reaction have to be removed in a revamped recycle-gas system, or by using a big amount of purge gas. In addition, the deoxygenation reactions had a tendency to compete with the primary desulfurization reactions taking place within the hydrotreating unit. This was seen to present too much risk in a fuel environment where ultra-low-sulfur diesel (ULSD) is required and rerun of off-spec products is expensive.

In view of the several difficulties related to co-processing route, building a dedicated unit optimized for vegetable-oil processing, as shown in figure, seems more cost effective.

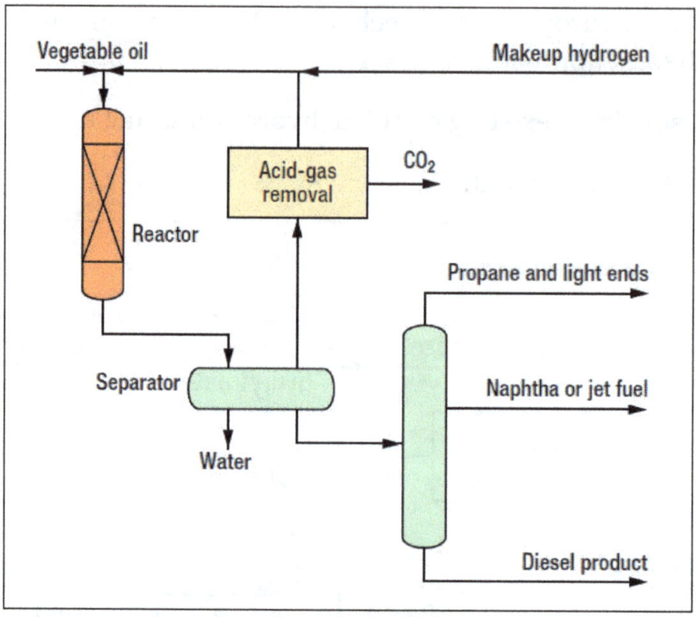

Green Diesel production flow scheme.

Figure reports a simplified flow scheme of the green diesel production process in a standalone unit. Vegetable oil is combined with hydrogen and brought to reaction temperature. Then the mixture is sent in the reaction unit where the vegetable oil is converted in Green Diesel. The last one is separated from the recycle gas in the separator unit while the liquid product is sent to a fractionation section. The design of the fractionation section can vary from a one-column system to produce diesel and unstabilized naphtha to a three-column system to produce propane, naphtha and diesel products. The recycle gas is treated in an amine system to remove CO_2.

Aviation Biofuel

Fuels like methanol and ethanol are not practical for aviation because they have very low energy densities. Planes would either be severely limited in their range or would not be able to take off thanks to the weight of the fuel they would need to carry. Aviation fuel has an energy density of 42 to 50 MG/kg, which is roughly the same as gasoline. In order for biofuels to compete with fossil fuels, they need to pack more punch.

Standard Aviation Fuel

To understand what an aviation biofuel needs to be, it is important to understand what makes current aviation fuels practical. This chart lists the major properties that are required of a fuel that will be used in planes and helicopters.

Important Properties of Aviation Fuels
High Quality
Does Not Freeze
Low Risk of Explosion
High Octane
Few Contaminants

The engines that are found in aircraft come in two types: turbines and piston engines. Each requires a different kind of fuel. The production of both of these fuels focuses on providing high power outputs and stable performance under the demands of flight. Of critical importance is water. Water in aviation fuel can freeze and cause lines to clog at higher altitudes. This is one of the reasons that alcohols, which tend to attract water, are not useful as aviation fuels. Cold weather performance is the most important factor in aviation fuel besides energy density.

Avgas

This is short for aviation gasoline and it is used in standard piston-engine aircraft (propellers). Its major difference from gasoline is that it is less prone to explosion, has a higher octane, and won't freeze at the low temperatures of higher altitudes. Avgas is generally being replaced by other aviation fuels. Avgas has an energy density of about 45 MG/kg.

Jet Fuel

Jet fuel is similar in many ways to kerosene and comes in two basic types: Jet A and Jet B. Jet A fuel is designed for use in turbine engines. It is designed to rest explosion by having an autoignition temperature of 210 degrees Celsius. Jet A is subdivided into Jet A and Jet A-1. The major difference between these fuels is freezing point. Jet A fuel freezes at -40 Celsius while Jet A-1 freezes at -47 Celsius. Jet A-1 is used much more commonly than Jet A. The energy densities of these two fuels are nearly identical at about 43 MJ/kg.

Jet B fuel is "lighter" than Jet A, which means it is more prone to evaporation and thus more prone to explosion. Its major use is for cold climates where its freezing point of -60 Celsius gives it a major advantage. Jet B has an energy density of about 43 MJ/kg.

Uses of Aviation Biofuel

Despite the challenges, aviation biofuels have seen some use starting in 2008. The first flight, which was undertaken by Virgin Atlantic, used a blend of 20% biofuels. This was followed by 50-50 blends through 2012. Then in October 2012, 100% biofuels was used by the National Research Council in Canada to power a Dassault Falcon 20.

Production of Aviation Biofuel

The aviation biofuels were no different, chemically, from standard fossil fuels. It is the case that direct alcohols cannot be used as aviation fuel because they freeze easily and have low energy densities. However, alcohols can be converted to kerosene, which is the basis for all aviation fuels.

Production of kerosene from biomass can occur in several different ways. Research into the use of biological organisms is ongoing and not yet viable. Current conversion processes take the form chemical cracking and gasification, which are energy intensive and do not represent viable solutions to the large-scale production of biofuels. At this point, aviation biofuel is more of a research curiosity than a practical consideration.

Biomass Feedstock

Where the biomass for producing aviation fuel comes from plays large part in how environmentally friendly these fuels are. Both the type of plant used and the location in which it is grown are important. Using arid or former agricultural land to produce biofuels feedstock can reduce greenhouse gas emissions. Plants like Jatropha or algae can be grown in these settings and are under investigation for use as feedstock. A study has shown, however, that using natural woodland to grown these plants (that is cutting down existing forest to create land for growing plants) will increase greenhouse gas emissions over the use of fossil fuels.

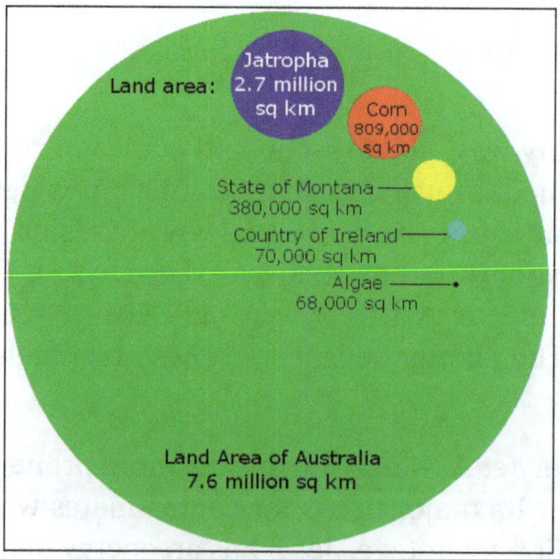

The graph above demonstrates something fantastic. Each year, 809,000 square kilometers of corn are planted, which is a land area roughly twice as large as the U.S. state of Montana. If Jatropha were used exclusively to create aviation fuel, 2.7 million square kilometers would be needed to meet current demand. Said another way, about 36% of the land area of Australia would be need to grow enough of the plant Jatropha. It is easy to see that this could have tremendous impact on land use and the food chain.

To get an idea of just how much land would be needed to meet current demands for aviation fuel, let's consider the following graph, which shows the land areas needed if a particular feedstock is to fully replace fossil fuels in aviation and compares those to well-known land masses.

Hydrogen Production using Microbial Biomass

Microbial biomass conversion processes take advantage of the ability of microorganisms to consume and digest biomass and release hydrogen. Depending on the pathway, this research could result in commercial-scale systems in the mid- to long-term timeframe that could be suitable for distributed, semi-central, or central hydrogen production scales, depending on the feedstock used.

How does it Work?

In fermentation-based systems, microorganisms, such as bacteria, break down organic matter to produce hydrogen. The organic matter can be refined sugars, raw biomass sources such as corn stover, and even wastewater. Because no light is required, these methods are sometimes called "dark fermentation" methods.

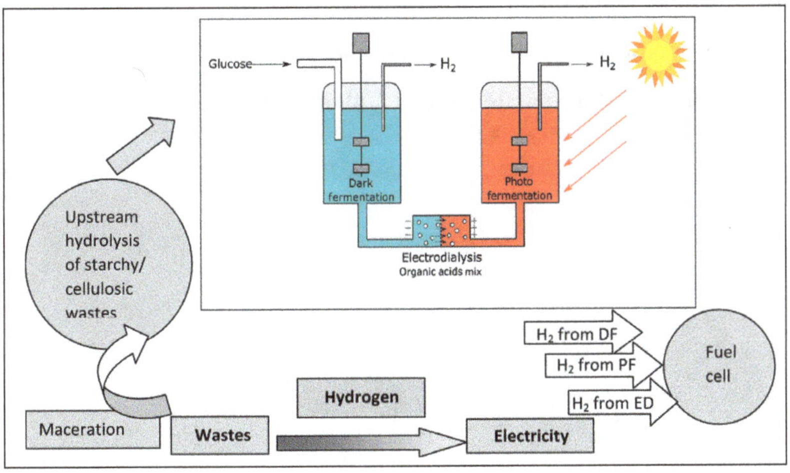

In direct hydrogen fermentation, the microbes produce the hydrogen themselves. These microbes can break down complex molecules through many different pathways, and the byproducts of some of the pathways can be combined by enzymes to produce hydrogen. Researchers are studying how to make fermentation systems produce hydrogen faster (improving the rate) and produce more hydrogen from the same amount of organic matter (increasing the yield).

Microbial electrolysis cells (MECs) are devices that harness the energy and protons produced by microbes breaking down organic matter, combined with an additional small electric current, to produce hydrogen. This technology is very new, and researchers are working on improving many aspects of the system, from finding lower-cost materials to identifying the most effective type of microbes to use.

Why is this Pathway being Considered?

Biomass is an abundant domestic resource, and many microbes have evolved to efficiently break down biomass to produce hydrogen and other products. Fermentation has already been used as an industrial technology to generate biofuels and other products, and many of the challenges to scaling up systems have been addressed for different products, allowing hydrogen researchers to focus on the challenges unique to hydrogen production. MEC-based systems have the potential to produce hydrogen from resources that otherwise can't be used for fuel production, and could reduce the large amount of energy normally needed for wastewater treatment while producing a valuable fuel in the form of hydrogen. These two pathways can be combined to maximize the hydrogen yield from the starting biomass feedstock.

References

- What-are-biofuels: ecomena.org, Retrieved 28, February 2020
- What-bioethanol: esru.strath.ac.uk, Retrieved 12, May 2020
- Bioethanol, biofuels: eubia.org, Retrieved 06, August 2020
- What-biodiesel, biofuels: esru.strath.ac.uk, Retrieved 10, June 2020
- Green-diesel: oil-gasportal.com, Retrieved 13, March 2020

Chapter 4

Conventional and Unconventional Biofuel Sources

Science and technology has undergone tremendous development in the past decade, which has resulted in the production of various biofuels. Some of the common sources of biofuels are corn, sugarcane, cellulose, Camelina, Jatropha, rapeseed, algal oil, etc. which have been extensively examined in this chapter.

Corn

Corn (Zea mays) is a popular feedstock for ethanol production in the United States due to its abundance and relative ease of conversion to ethyl alcohol (ethanol). Corn and other high-starch grains have been converted into ethanol for thousands of years, yet only in the past century has its use as fuel greatly expanded. Conversion includes grinding, cooking with enzymes, fermentation with yeast, and distillation to remove water. For fuel ethanol, two more steps are included: using a molecular sieve to remove the last of the water and denaturing to make the ethanol undrinkable.

Current Potential for use as a Biofuel

Corn grain makes a good biofuel feedstock due to its starch content and its comparatively easy conversion to ethanol. Infrastructure to plant, harvest, and store corn in mass quantities benefits the corn ethanol industry. Unlike sugarcane, in which squeezed sugar water can be directly fermented, corn starch must be cooked with alpha and gluco-amylase enzymes to convert the starch to simple sugars. Cellulosic feedstocks are even more recalcitrant and require time and energy to convert to simple sugars. Under the renewable fuel standard set by Congress in 2007 (RFS-2), grain-based ethanol can make up 15 billion gallons of the 36 billion gallon-per-year requirement. Corn-based ethanol production capacity in 2009 was 10.6 billion gallons. The addition of idled capacity would increase potential production to 12.5 billion gallons per year (Renewable Fuels Association).

Corn production in the United States reached record highs in 2009 with 13.2 billion bushels from 86.5 million acres (National Ag Statistics Service). Using the current corn-to-ethanol conversion of 2.8 gallons of ethanol from a bushel of corn, total U.S. corn production could result in approximately 37 billion gallons of ethanol, which would provide approximately 26% of our 137 billion gallon-per-year gasoline consumption (Energy Information Administration). However, using all of our corn for ethanol is neither realistic nor necessary and has not been proposed. Creating the 15 billion gallons required under the RFS-2 would call for 5.4 billion bushels or about 41% of our 2009 corn crop. Although this percentage seems rather high, one-third of the weight and 100% of the nutritional content of corn entering an ethanol dry mill biorefinery is returned to the feed market as distillers grains. These distillers grains can be used to replace corn in the diets of cattle, swine, and poultry. When this replacement is calculated into the overall consumption figures, it lowers the number to 27% of our 2009 corn crop, or only 3.6 billion bushels of corn to produce the 15 billion required gallons. Throughout history, the United States has seen a steady increase in the yields of both corn and ethanol production. It is very likely that the United States will be able to increase corn ethanol production without expanding to new acres and still have plenty of corn remaining to meet other domestic use and export demands.

One bushel of corn grain will yield one-third ethanol, one-third distillers grains, and one-third carbon dioxide, or 17 pounds of distillers grains and 2.8 gallons of ethanol.

Biology and Adaptation

Corn (Zea mays) originated in Central America with the first domestication, purported to be in the Tehuacan Valley of Mexico. Spreading throughout the North American continent, corn became an important crop for early Americans. At its peak in 1917, 111 million acres of corn were planted in the United States. Today corn is planted on every continent in the world except Antarctica and is grown throughout many states in the United States, ranging from southern North Dakota to Texas and eastward to New York. Corn is well adapted to growing in temperatures between 50° and 86°Fahrenheit. To produce grain, corn will use approximately 22 to 28 inches of water, which requires 12 to 20 inches of rainfall or irrigation during the growing season. Despite popular misconceptions, nearly 90% of U.S.-grown corn is fed by natural rainfall only, with no irrigation necessary. Many parts of the upper Midwest are well suited to grow corn, and this area is sometimes referred to as the Corn Belt.

Production and Agronomic Information

Corn in the upper Midwest is seeded between March and May and harvested between

September and November in most years. A majority of corn planted today has genetic resistance to some weeds, insects, and plant pathogens. Corn hybrid resistance to various pests and pathogens is a result of biotechnology and plant breeding. Biotechnology traits aid producers in the control of weeds and insects, greatly reducing the amount of pesticides entering the environment. Much of the Corn Belt rotates with other crops, such as soybeans or wheat, to break weed, insect, and disease cycles as well as to reduce the cost of production. Corn responds best to highly fertile soils with supplemental fertilizer applied in most years. Fertilizer may be inorganic chemical fertilizer or manure.

Major nutrients required by corn are nitrogen, phosphorus, and potassium. Inorganic nitrogen fertilizer production is very energy-intensive and as a result, nitrogen fertilizer represents nearly 30% of the energy inputs in corn production. Other major inputs include diesel fuel for tractors, transportation, irrigation, and electricity for irrigation and grain storage.

Potential Yields

The average national corn yield was 165 bushels per acre in 2009. Corn yields have increased by approximately 2 bushels per acre each year since 1940. This increase will likely continue into the future, with some predicting the yield trend will amplify at a greater rate due to biotechnology and advancements in breeding. Ethanol yield per acre would be 462 gallons per acre from corn yields of 165 bushels per acre. An acre of sugarcane can produce an approximate 35 ton yield, resulting in about 560 gallons of sugarcane ethanol.

Production Challenges

Corn production has been blessed with nearly 100 years of infrastructure buildup and research. Farmers have great knowledge and experience in growing corn. This infrastructure and grower intelligence make corn a natural crop for expanded uses such as ethanol. Yet high production costs and high inputs make corn a very intensive crop. Other bioenergy crops may be less intensive and require fewer inputs. The cost versus profit per acre needs to be compared, as economics is a major driver in deciding which crop is best. Growing another crop on an acre where corn could be grown carries risks that may include a new cropping system; no harvest, transport, or storage infrastructure; or no commodity market to fall back on if the biofuel market fails.

Estimated Production Costs

Production costs vary widely depending on tillage, irrigation, yield goal (soil fertility), spraying schedule, seed selection, and rotation. A sample corn budget with rain-fed, no-till, biotech seed, corn/soybean rotation, and 120 bushel yield goal would include a total cost of $211 per acre. If overhead – crop insurance, land, taxes – is included, the total is $305 per acre. Production costs increase to over $600 on irrigated fields with continuous corn.

Environmental and Sustainability Issues

Life cycle analysis (LCA) of ethanol production from corn grain has yielded a net energy ratio of 1.2 to 1.45, which represents just a 20% to 45% positive energy balance in producing ethanol from corn. A major criticism of corn ethanol has been the large amount of fossil energy used in production.

Environmental issues in corn production revolve around erosion, pesticide use, and nutrient use. Pesticides and nutrients have the potential to contaminate surface and ground water. Soil erosion has led to loss of topsoil and polluted streams and river systems with silt. Continuous attention to these issues has led to improvements, yet they will remain concerns in crop production.

Sugarcane

Sugarcane is one of the most promising agricultural sources of biomass energy in the world. It is the most appropriate agricultural energy crop in most sugarcane producing countries due to its resistance to cyclonic winds, drought, pests and diseases, and its geographically widespread cultivation. Due to its high energy-to-volume ratio, it is considered one of nature's most effective storage devices for solar energy and the most economically significant energy crop. The climatic and physiological factors that limit its cultivation to tropical and sub-tropical regions have resulted in its concentration in developing countries, and this, in turn, gives these countries a particular role in the world's transition to sustainable use of natural resources.

Sugarcane is a highly efficient converter of solar energy, and has the highest energy-to-volume ratio among energy crops. Indeed, it gives the highest annual yield of biomass of all species. Roughly, 1 ton of Sugarcane biomass-based on Bagasse, foliage and ethanol output – has an energy content equivalent to one barrel of crude oil. Sugarcane produces mainly two types of biomass, Cane Trash and Bagasse. Cane Trash is the field residue remaining after harvesting the Cane stalk and Bagasse is the milling by-product which remains after extracting sugar from the stalk. The potential energy value of these residues has traditionally been ignored by policy-makers and masses in developing countries. However, with rising fossil fuel prices and dwindling firewood supplies, this material is increasingly viewed as a valuable renewable energy resource.

Sugar mills have been using Bagasse to generate steam and electricity for internal plant requirements while Cane Trash remains underutilized to a great extent. Cane Trash and Bagasse are produced during the harvesting and milling process of Sugarcane which normally lasts 6 to 7 months.

Around the world, a portion of the Cane Trash is collected for sale to feed mills, while freshly cut green tops are sometimes collected for farm animals. In most cases, however, the residues are burned or left in the fields to decompose. Cane Trash, consisting of Sugarcane tops and leaves can potentially be converted into around 1kWh/kg, but is mostly burned in the field due to its bulkiness and its related high cost for collection/transportation.

On the other hand, Bagasse has been traditionally used as a fuel in the Sugar mill itself, to produce steam for the process and electricity for its own use. In general, for every ton of Sugarcane processed in the mill, around 190 kg Bagasse is produced. Low pressure boilers and low efficiency steam turbines are commonly used in developing countries. It would be a good business proposition to upgrade the present cogeneration systems to highly efficient, high pressure systems with higher capacities to ensure utilization of surplus Bagasse.

Sugarcane Trash as Biomass Resource

Sugarcane trash (or cane trash) is an excellent biomass resource in sugar-producing countries worldwide. The amount of cane trash produced depends on the plant variety, age of the crop at harvest and soil and weather conditions. Typically it represents about 15% of the total above ground biomass at harvest which is equivalent to about 10-15 tons per hectare of dry matter. During the harvesting operation around 70-80% of the cane trash is left in the field with 20-30% taken to the mill together with the sugarcane stalks as extraneous matter.

Cane trash's calorific value is similar to that of bagasse but has an advantage of having lower moisture content, and hence dries more quickly. Nowadays only a small quantity of this biomass is used as fuel, mixed with bagasse or by itself, at the sugar mill. The rest is burned in the vicinity of the dry cleaning installation, creating a pollution problem in sugar-producing nations.

Cane trash and bagasse are produced during the harvesting and milling process of sugarcane which normally lasts between 6 to 7 months. Cane trash can potentially be converted into heat and electrical energy. However, most of the trash is burned in the field due to its bulky nature and high cost incurred in collection and transportation.

Cane trash could be used as an off-season fuel for year-round power generation at sugar mills. There is also a high demand for biomass as a boiler fuel during the sugar-milling season. Sugarcane trash can also converted in biomass pellets and used in dedicated biomass power stations or co-firedwith coal in power plants and cement kilns.

Burning of cane trash creates pollution in sugar-producing countries.

Currently, a significant percentage of energy used for boilers in sugarcane processing is provided by imported bunker oil. Overall, the economic, environmental, and social implications of utilizing cane trash in the final crop year as a substitute for bunker oil appears promising. It represents an opportunity for developing biomass energy use in the Sugarcane industry as well as for industries / communities in the vicinity.

Positive socio-economic impacts include the provision of large-scale rural employment and the minimization of oil imports. It can also develop the expertise necessary to create a reliable biomass supply for year-round power generation.

Recovery of Cane Trash

Recovery of cane trash implies a change from traditional harvesting methods; which normally consists of destroying the trash by setting huge areas of sugarcane fields ablaze prior to the harvest. There are a number of major technical and economic issues that need to be overcome to utilize cane trash as a renewable energy resource. For example, its recovery from the field and transportation to the mill, are major issues.

Alternatives include the current situation where the cane is separated from the trash by the harvester and the two are transported to the mill separately, to the harvesting of the whole crop with separation of the cane and the trash carried out at the mill. Where the trash is collected from the field it maybe baled incurring a range of costs associated with bale handling, transportation and storage. Baling also leaves about 10-20% (1-2 tons per hectare) of the recoverable trash in the field.

A second alternative is for the cane trash to be shredded and collected separately from the cane during the harvesting process. The development of such a harvester-mounted cane trash shredder and collection system has been achieved but the economics of this approach require evaluation. A third alternative is to harvest the sugarcane crop completely which would require an adequate collection, transport and storage system in addition to a mill based cleaning plant to separate the cane from the trash.

A widespread method for cane trash recovery is to cut the cane, chop into pieces and then it is blown in two stages in the harvester to remove the trash. The amount of trash that goes along with the cane is a function of the cleaning efficiency of the harvester. The blowers are adjusted to get adequate cleaning with a bearable cane loss.

On the average 68 % of the trash is blown out of the harvester, and stays on the ground, and 32 % is taken to the mill together with the cane as extraneous matter. The technique used to recover the trash staying on the ground is baling. Several baling machines have been tested with small, large, round and square bales. Cane trash can be considered as a viable fuel supplementary to bagasse to permit year-round power generation in sugar mills.

Thus, recovery of cane trash in developing nations of Asia, Africa and Latin America implies a change from traditional harvesting methods, which normally consists of destroying the trash by setting huge areas of cane fields ablaze prior to the harvest. To recover the trash, a new so-called "green mechanical harvesting" scheme will have to be introduced. By recovering the trash in this manner, the production of local air pollutants, as well as greenhouse gases contributing to adverse climatic change, from the fires are avoided and cane trash could be used as a means of regional sustainable development.

Soybean

Soybean oil is a major feedstock for biodiesel production. Soybeans contain approximately 18% oil.

Soybean as a Biofuel Feedstock

Soybean (Glycine max) is a major crop throughout much of North America, South America and Asia. The United States is the world's greatest producer, producing approximately 32% of the worlds soybeans followed by Brazil with 28%. Origins of soybean are in Southeast Asia with first domestication reported in the 11th century BC in

China. First planted in the U.S. in 1765, soybeans spread to the Corn Belt by the mid 1800s with major acreage not seen until the 1920s when it was used mainly as a forage crop. Major U.S. expansion as an oilseed crop began in the 1940s. Soybeans contain approximately 18 to 20% oil compared to other oilseed crops such as canola (40%) and sunflower (43%). At 48 lbs per bushel soybean meal remains a major product from soybeans used for animal feed and human food. Soybean acreage is much greater than other oilseed crops leading to substantial soybean oil production and its availability as a biofuel feedstock.

Current Potential for Use as a Biofuel

Soybean oil is currently a major feedstock for production of biodiesel (NBB). The most common method of biodiesel production is a reaction of vegetable oils or animal fats with methanol or ethanol in the presence of sodium hydroxide (which acts as a catalyst). The transesterification reaction yields methyl or ethyl esters (biodiesel) and a byproduct of glycerin. Note biodiesel is not straight vegetable oil burned in a diesel engine. Numerous studies between 1980 and 2000 have shown the use of straight vegetable oil including soybean oil to cause carbon deposits and shorten engine life. Biodiesel use in diesel engines does not have similar negative effects. Use of soybean oil for biodiesel was greatly influenced by promotion from U.S. soybean farmers through the United Soybean Board (USB) and subsequent creation of the National Biodiesel Board (NBB).

Biology and Adaptation

Soybean (Glycine max) is a cool season legume which can be grown from south to north throughout much of the eastern half of the U.S. Soybeans and other legumes have a unique relationship with a bacteria bradyrhizobium species, will colonize on soybean roots forming a nodule. The two species form a symbiotic relationship where the soybean plant provides nutrition and the bacteria fixes nitrogen from the air. This relationship reduces the need for supplemental nitrogen fertilizer in soybean production.

Soybeans flower in response to day length and temperature. Varieties grown in the United States are divided into 13 maturity groups from maturity group 000 being the earliest and adapted to northern regions of Minnesota and southern Canada, to maturity group X adapted to southern regions such as south Texas. The earlier varieties bloom when days are long and nights are short, while the later-maturing varieties bloom under relatively shorter days and longer nights. Summer days are longer at northern latitudes, where early maturing varieties will initiate flowering when days are longer. maturity groups develop differently and knowing the growth habit of different maturity groups can help with the crop management.

Production and Agronomic Information

Through much of the upper Midwest soybeans are planted in April to June and harvested in September to November. Soybeans are well adapted to grow in soils similar to corn production. In many cases soybeans are grown in rotation with corn or wheat to break insect, weed, and disease cycles. Nutrient requirements are generally less for soybeans than other crops with major nutrient requirements nitrogen, phosphorous, and potassium and where much of the nitrogen is gained through a relationship with bacteria. A soil pH in the range of 5.5 to 7.0 will enhance nutrient availability and soybean growth. Weed control is necessary to achieve optimal yields and use of biotech seeds has eased the ability to control weeds during the growing season. Currently in the U.S. over 90% of soybeans planted are herbicide resistant. Many insects and diseases are common in soybeans grown in the upper Midwest. The most damaging pest to soybeans is soybean cyst nematode, a soil born parasitic roundworm feeding on soybean roots. Insect pests include: bean leaf beetle, soybean aphid, green clover worm, and spider mites. Soybean harvest begins after 95% leaf senescence when beans are 12 to 18% moisture.

Potential Biofuel Yields

Current U.S. production of soybeans in 2009 was 3.4 billion bushels from 77.4 million acres. Average yield per acre for the U.S. was 44 bushels per acre (National Ag Statistics Service). One bushel of soybeans can yield 1.5 gallons of biodiesel (NBB). Using all U.S. soybeans for biodiesel could produce 5.1 billion gallons biofuel. Using all soybean production for biodiesel has not been proposed and is not realistic. In 2009 biodiesel production was 700 million gallons with a production capacity of 1.83 billion gallons. Based on a yield of 44 bushels per acre, an acre of soybeans could yield 66 gallons of biodiesel compared to 69 gallons for a 1300 lb per acre canola yield, 84 gallons for sunflower and over 600 gallons for palm oil.

Production Challenges

Soybean production generally compliments corn production in the upper Midwest. Both corn and soybeans enjoy a long history of production on millions of acres in the upper Midwest. This history has led to a large infrastructure of equipment, storage, rail, barge and truck transportation. Soybeans like many crops face insect and disease pests along with weather related challenges. An emerging disease has gained much attention in recent years. Soybean rust, a fungal disease native to Asia has spread to the soybean fields of South America and finally to U.S. soybeans. Rust control is expensive requiring fungicide applications and yield damage can be extreme.

Estimated Production Costs

Production costs will vary depending on location, cropping systems, and fluctuation in price of energy. Major expenditures in soybean production include; Planting, harvesting, seed and pesticides. An example of a Nebraska rainfed budget for no-till soybeans for 2010 lists $115 per acre for field operations, materials and services. When including overhead costs for land, insurance, etc. the total is approximately $200 per acre. Total costs for irrigated soybeans are around $400 per acre. Biodiesel profitability is extremely variable and based on the continuously changing prices biodiesel, soybean oil, co-product glycerin, methanol and natural gas. Price of soybean oil feedstock is one of the driving factors in profitability.

Environmental and Sustainability Issues

Soybeans use for biodiesel production capacity grew from zero to over a billion gallons per year in the past two decades 90s and 00s. Through that time biodiesel production rose and fell depending on the price of feedstock, price of petroleum oil, and federal and state subsidies provided to the industry. One major challenge for soybeans are competing uses for soybean oil. Soybean oil is used for many human food products, cooking oil and numerous industrial applications. Soybeans account for 80 percent or more of the edible fats and oils consumed in the U.S. Competition with other uses has caused price spikes in the soybean oil market challenging the profitability of soybean biodiesel. A life cycle analysis of biodiesel done by the USDA reports the fossil energy ratio of biodiesel to be 3.2 units of energy out for each unit of fossil energy used. In other words, biodiesel yields 3.2 units of energy for every unit of fossil energy consumed compared to petroleum diesel which has a fossil energy ratio of ~0.84.

Vegetable Oil

Vegetable oil can be used to fuel diesel engines as straight vegetable oil (SVO) or as biodiesel following conversion. It can be used either for cars, cooking or for stationary use to power machinery or an electricity generator. The more advanced the engine technology, the more complicated or limited the use of SVO. Hence for use in modern cars with direct injection engines the conversion of SVO to biodiesel is mandatory. However SVO can be directly used as fuel for most simple stationary diesel engines used in generator sets or pump stations. Special SVO engines are also available. For robust stationary (non-vehicle) engines, the use of pure SVO is the most attractive option. This way the additional and somewhat dangerous biodiesel conversion process (due to some highly inflammable components) can be avoided; furthermore the energy yield per gallon of vegetable oil is higher when using SVO directly.

Listed below are the potential benefits of SVOs as fuel for power generation in rural areas:

- SVOs are a renewable energy source low in greenhouse gas (GHG) emissions.
- Local production is possible and can contribute to value generation in rural areas.

- SVOs are liquid, hardly evaporate and are thus easily handled, stored and transported.
- SVOs are neither flammable nor explosive and do not release toxic gases.
- SVOs can be burnt in a relatively clean manner.
- SVOs do not cause major damage if accidentally spilt.

Almost all vegetable oils have a calorific value that is very similar to that of diesel fuel. However vegetable oil differs from diesel considerably with regard to viscosity, which is 10–20 times higher for SVOs. Furthermore, the solidification points of SVOs can be rather high. Coconut oil for example solidifies at temperatures below 22°C. This can cause major prob-ems particularly in cold regions. Oil characteristics can differ considerably between different types of SVOs as their composition of fatty acids varies. This can lead to plugging and gumming of lines, filters and injectors, as well as cause deposits inside the motor or excessive engine wear. There is long-term experience particularly with rapeseed oil, which has been used in Germany and other countries as fuel for cars, trucks and tractor engines for several years. There is no doubt that rapeseed oil is relatively easy and reliable to use in diesel engines. Other oils like palm oil, sunflower, soybean, coconut, jatropha and cotton oil have also been used to fuel motors. Many examples are regularly presented at conferences and can be found in publications. However there does not yet seem to be much large-scale and long-term practical experience regarding the proper and economic long-term functioning of such plant oil motor-generator systems. For economic long-term use all machines need clean homogenous oil with specific characteristics.

Sources of Vegetable Oil

Man currently uses over one thousand different oil plants. Some of their seeds are extracted from natural forests; some are cultivated in small plots or hedges, while others can be pro-duced in plantations with intensive agricultural methods. The most cultivated oil plants of the tropics are oil palm, coconut, soy, castor, sunflower and cotton. In many cases oil is only a by-product that is rarely extracted.

In the discussion on biofuels jatropha received particular focus due to its attributes:

- Wide adaptability to diverse climatic zones and soil types combined with high drought tolerance.
- Cultivable in areas where food production is not possible, therefore not competing with food crops.
- Short gestation period, easy multiplication and high pest and disease resistance.

The oil from cottonseed can be used as an energy source. Cotton oil can also be used for human consumption. However it is of minor value and due to the intensive use of chemical pest control in cotton culture its agrochemical content can be problematic for alimentation. In cases where cotton oil is available at reasonable prices it might serve as a sound source of energy. Usually cotton oil is not extracted as the cottonseed is used as forage for domestic animals.

Palm oil is produced at both industrial and small-scale levels. The oil palm is one of the most productive oil plants in terms of produced oil per hectare. Coconut and sunflower are also important sources of vegetable oil. All these oils have high value for human consumption and are therefore preferably used for nutritional purposes. However, in some particular circum-stances the energetic use of these oils might prove reasonable.

There are considerable advantages for a farmer if he has several options to market his oil produce. So in cases of low quality oil production or very low prices for edible oil he can sell or use it for energy purposes.

There are many other agricultural plants cultivated for food, fodder or fibre that contain oil. Some wild or semi-wild growing trees are also occasionally used as oil sources. In some cases this oil can be available at costs and in quantities that allow its use as fuel. The use of kapok oil in Cambodia is an example.

Besides fresh vegetable oil, waste oils - for example from restaurants - can also be used as a fuel source. Almost all countries (should) have regulatory frameworks that inhibit irregular disposal of oils and fats. Hence the recycled fraction of these materials is high. Much of this is processed at industrial level. But examples of small-scale oil and biodiesel productions do exist.

Extraction

The two main processes of extracting oil from seed feedstock are mechanical press extrac-tion and solvent extraction.

The solvent process extracts more of the oil contained in the oilseed feedstock but requires costlier equipment. Oil extraction by a mechanical press is the standard technology for small-scale applications. During mechanical press extraction, the oilseed is crushed and pressed in a screw press. The oilseed is heated either before or during the pressing. After most of the oil is removed, the remaining seed meal can be used as animal feed.

Hand presses are not appropriate for extraction of larger seed quantities. Screw presses driven by combustion engines are standard and widely available in the market. The efficiency and durability of the different available presses will probably be a topic for future research and development.

After the crushing and pressing, the oil has to be filtered. Chamber filters with fine mesh tex-tile filters are the most common option.

Market Development

The developments in the world biofuel market are important for any local approach. Many jatropha and other oil plant plantations have been established due to the biofuel hype of the recent years. Even though their yields do not fulfil the original expectations, these plantations will produce considerable amounts of oilseeds within the next few years. This will probably lead to an increasing availability of SVO on the markets; SVO production know-how will grow considerably with the lessons learnt from cultivation during this time.

And the gold rush mood is not yet over. New crops and plants are being touted as 'new oil wells,' e.g. the Acrocomia palm tree in Southern America was recently promoted as an in-sider tip. The implications of this gold rush mood are not yet clearly visible.

The demand for vegetable oils as a base for car fuels will persist and increase on the global market. There are some sectors like the aviation business with high purchasing power and technical applications that have no alternative to liquid fuels. Their demand for liquid fuels will therefore remain or increase further.

The development of environmental and social standards will most likely ensure that in future only certified SVO or biodiesel will find its way into the formal market. This will probably lead to high competition and low availability of cheap SVOs in the local markets of developing countries. However, some quantities of SVOs that do not meet international standards might remain for local use.

In any case the development of the global markets will increase infrastructure and know-how about SVO production and use, including high yield varieties, agricultural extension service, availability of appropriate engines etc. Local initiatives will probably also benefit from this development.

Cellulose

Cellulose is a polymer or string of glucose molecules joined together in linear rows. The parallel rows of glucose molecules form a tough crystalline substance which gives wood, paper and cardboard their strength. Cellulose comprises about 40-60% of the material in common forms of biomass, such as wood, paper, switchgrass, and corn stover. Cellulose is probably the most abundant organic molecule on the planet.

The other biomass components are hemicellulose, made from 5-carbon sugars like xylose, and lignin, which is a set of non-sugar molecules acting like a glue to hold the biomass molecules together. The 5-carbon sugars, or pentoses, can also be fermented to ethanol, and the lignin can be burned for energy or spun into high-strength carbon fibers for the light carbon composites used in wind turbine blades and aircraft parts. With further cost reductions, carbon fibers made from biomass lignin should become part of an increasing new market for light weight automobiles, including EVs. Glucose, xylose and lignin from biomass are currently the only feasible alternatives for large-scale substitution of petroleum hydrocarbons.

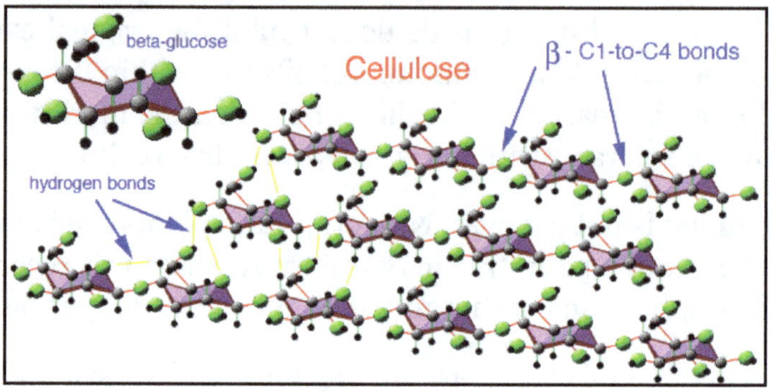

Cellulosic ethanol is fuel ethanol made from glucose, a 6-carbon sugar derived from the cellulose in biomass. Cellulosic ethanol is a substitute for gasoline. It is chemically identical to ethanol made from food crops like corn and sugar, but comes from wood, waste paper, and energy crops like poplar and switchgrass. Cellulosic ethanol is more difficult to make, because cellulose is a tough structural material, unlike starch from grains which is easily broken down to glucose sugar.

The primary feedstocks for fuel ethanol today include glucose from corn starch in the U.S., currently producing about 12 billion gallons of corn ethanol per year. Additional corn ethanol plants could push the total to 15 billion gal/year, with the downside of increasing corn prices and encouraging the shift of wheat and soybean acreage to corn. One bushel of corn produces about 2.8 gal ethanol, so 12 billion gal corn ethanol requires about 4.3 billion bu/year, a significant fraction of the U.S. corn crop.

In Brazil, the primary feedstock for fuel ethanol is sucrose (white sugar) from sugar cane, a more efficient source than corn. Standard gasoline in Brazil is about 25% ethanol, and flexfuel vehicles using E85 are popular. Growing large amounts of sugar cane is not an option in the U.S., Canada and Europe because of the tropical climate required.

Cellulosic ethanol can be made from a variety of biomass feedstocks including recycled paper, urban waste paper diverted from municipal solid waste (MSW), agricultural wastes like sugarcane bagasse and corn stover, energy crops like poplar, willow, and switchgrass, wood waste, and waste streams from pulp and paper mills. Cellulose and hemicellulose are the two carbohydrate components in most biomass, and can be hydrolyzed (split) into their component sugars: glucose, xylose, both of which can be fermented to ethanol, other biofuels like butanol, green diesel or biojetfuel, or lactic acid which can be made into biodegradable or more permanent plastics and fabrics. Cellulosic ethanol is also a sustainable feedstock for drop-in bioplastics like renewable plant-based polyethylene and polypropylene.

Cellulosic ethanol made from biomass will produce smaller amounts of greenhouse gases than corn ethanol, and far less than the gasoline it will replace. This greenhouse reduction potential can be estimated by Life Cycle Analysis (LCA), used by numerous groups to evaluate the net energy balance and carbon balance of ethanol from corn and cellulosic biomass.

The technologies can be broadly categorized in two path ways:

1. Biochemical conversion (fermentation) through pretreatment and hydrolysis.
2. Thermo-chemical conversion through gasification.

Biochemical Conversion

The first step is to break down cellulose which requires pretreatment. In pretreatment process, hemicelluloses and lignin that surround cellulose are broken down under a moderately high-temperature, high-pressure and through use of dilute acid. This process which is called hydrolysis breaks down hemicelluloses and dissolves lignin. The lignin thus produced forms and important source of heat and electricity, hence, limiting the use of fossil fuel in the conversion process. However, the problem with lignin is that it can under certain conditions during the pretreatment, redeposit onto cellulose which ultimately reduces the yield of sugar. Also this dilute acid treatment makes the process expensive as it requires costly equipment.

It has been proven that a milder pretreatment process whereby an appropriate mixture of enzyme to the hydrolysis of hemicellulose can curb degradation of sugar and ultimately the cost of processing.

Another way of enhancing the pretreatment process has been identified as the Ammonia Fiber Explosion (AFEX) process in which the lignocellulosic biomass is treated with high-pressure liquid ammonia leading to the explosive release of the pressure and thereby rendering the lignocellulosic biomass more susceptible to the enzymatic hydrolysis.

The simple sugars broken down from the cellulosic materials are fermented using yeast or bacteria under ideal conditions. These microorganisms convert the sugar into ethanol and water which is called the ethanol recovery process. The water is removed through distillation. The process is similar to that of bioethanol.

Thermo-chemical Conversion

In thermo-chemical conversion of cellulose into ethanol, the ligno-cellulosic raw material is broken into syngas i.e., carbon monoxide and hydrogen first applying heat and chemicals. This process is mostly appropriate when forest products and mill residues that are rich in lignin are used as feedstock and which cannot be converted by biochemical process. This process is however, complex and is similar to that of petrol refining in which contaminants (tar, sulphur, etc.) are also produced along with the syngas. The syngas are converted into ethanol which then undergoes distillation.

Feasibility of Technology and Operational Necessities

Production of cellulosic ethanol for large scale commercial use also requires cultivation of energy crops besides collection of waste as feedstocks. Therefore this requires adequate land, suitable soil and adequate water. Recent studies suggest of world's 13.5 billion hectares of surface lands, approximately 1.6 billion hectares of land is used as cropland. Therefore, availability of land besides forest land, protected areas and cultivated food cropland that could be used for producing energy crops is estimated between 250-800 million hectares and most of which lies in the tropical regions of Latin America and Africa.

The very first logistics for producing ethanol from cellulose requires producing biomass which can be obtained from forest resources and/or agricultural resources. The produced biomass has to be harvested and then stored before it is finally transported to the conversion plant. These steps comprise theupstream logistics.

Before the biomass is introduced to the bio-refinery, the size in which they've been harvested needs to be reduced to the level where it is easy to handle and the process becomes more efficient. So if it is agricultural residue they need to be grinded and if wooden residue then these has to be taken through the chipping process so that the size of the materials are uniform.

Some of the key barriers to the commercialization of cellulosic ethanol as as under:

- Infrastructure barriers: Infrastructure for the distribution of cellulosic ethanol might be a barrier because it cannot be transported through pipes but has to be carried through vehicles by road. Georgraphical locations suitable for production of feedstock and the area of demand for the ethanol may vary thus requiring extensive investment in infrastructure development like road/railway networks. Also greater investment required for storing and processing especially in the pre-treatment stage.
- Market barriers: Lack of distribution infrastructure for market penetration, availability of a limited number of vehicle models that can run with the given type of fuel combination that uses high ethanol blend percentage and lack of awareness about the benefits of the cellulosic ethanol among the industry participants as well as the consumers are barriers related with the market place.
- Feedstock logistical and technical barriers: At present there is lack of efficiency in meeting the requirements related with sustainable harvesting of the feedstock and development of bio-refineries, harvest equipment and technology which can deal with the large yield of biomass. Production of cellulosic ethanol requires adequate land, water and suitable soil.
- Financial barriers: Expensive process and requires costly equipment, a high cost of cellulose enzyme is a barrier for economic production.

Algal Oil

Generally, algae are a diverse group of prokaryotic and eukaryotic organisms ranging from unicellular genera such as Chlorella and diatoms to multicellular forms such as the giant kelp, a large brown alga that may grow up to 50 m in length. Algae can either be autotrophic or heterotrophic. The autotrophic algae require only inorganic compounds such as CO_2, salts, and a light energy source for their growth, while the heterotrophs are non-photosynthetic, which require an external source of organic compounds as well as nutrients as energy sources. Microalgae are very small in sizes usually measured in micrometers, which normally grow in water bodies or ponds. Microalgae contain more lipids than macroalgae and have the faster growth in nature.

The algae can be converted into various types of renewable biofuels including bioethanol, biodiesel, biogas, photobiologically produced biohydrogen, and further processing for bio-oil and syngas production through liquefaction and gasification, respectively. The conversion technologies for utilizing algal biomass to energy sources can be categorized into three different ways, i.e., biochemical, chemical, and thermochemical conversion and make an algal biorefinery, which has been depicted in figure below.

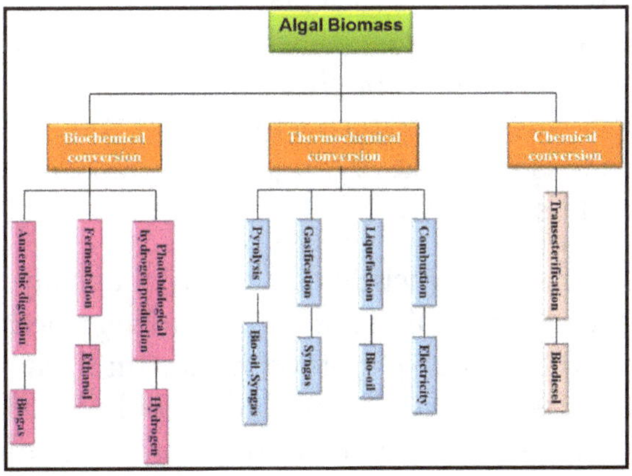

Algal biomass conversion process for biofuel production.

Algal biodiesel production involves biomass harvesting, drying, oil extraction, and further transesterification of oil, which have been described as below.

Harvesting and Drying of Algal Biomass

Unicellular microalgae produce a cell wall containing lipids and fatty acids, which differ them from higher animals and plants. Harvesting of algal biomass and further drying is important prior to mechanical and solvent extraction for the recovery of oil. Macroalgae can be harvested using nets, which require less energy while microalgae can be harvested by some conventional processes, which include filtration flocculation, centrifugation, foam fractionation, sedimentation, froth floatation, and ultrasonic separation. Selection of harvesting method depends on the type of algal species.

Drying is an important method to extend shelf-life of algal biomass before storage, which avoids post-harvest spoilage. Most of the efficient drying methods like spray-drying, drum-drying, freeze drying or lyophilization, and sun-drying have been applied on microalgal biomass. Sun-drying is not considered as a very effective method due to presence of high water content in the biomass. However, Prakash et al. used simple solar drying device and succeed in drying Spirulina and Scenedesmus having 90% of moisture content. Widjaja et al. showed the effectiveness of drying temperature during lipid extraction of algal biomass, which affects both concentration of triglycerides and lipid yield. Further, all these processes possess safety and health issues.

Extraction of Oil from Algal Biomass

Unicellular microalgae produce a cell wall containing lipids and fatty acids, which differ them from higher animals and plants. In the literature, there are different methods of oil extraction from algae, such as mechanical and solvent extraction. However, the extraction of lipids from microalgae is costly and energy intensive process.

Mechanical oil extraction. The oil from nuts and seeds is extracted mechanically using presses or expellers, which can also be used for microalgae. The algal biomass should be dried prior to this process. The cells are just broken down with a press to leach out the oil. About 75% of oil can be recovered through this method and no special skill is required. Topare et al. extracted oil through screw expeller by mechanical pressing (by piston) and osmotic shock method and recovered about 75% of oil from the algae. However, more extraction time is required as compared to other methods, which make the process unfavorable and less effective.

Solvent based oil extraction. Oil extraction using solvent usually recovers almost all the oil leaving only 0.5–0.7% residual oil in the biomass. Therefore, the solvent extraction method has been found to be suitable method rather than the mechanical extraction of oil and fats. Solvent extraction is another method of lipid extraction from microalgae, which involves two stage solvent extraction systems. The amount of lipid extracted from microalgal biomass and further yield of highest biodiesel depends mainly on the solvent used. Several organic solvents such as chloroform, hexane, cyclo-hexane, acetone, and benzene are used either solely or in mixed form. The solvent reacts on algal cells releasing oil, which is recovered from the aqueous medium. This occurs due to the nature of higher solubility of oil in organic solvents rather than water. Further, the oil can be separated from the solvent extract. The solvent can be recycled for next extraction. Out of different organic solvents, hexane is found to be most effective due to its low toxicity and cost.

In case of using mixed solvents for oil extraction, a known quantity of the solvent mixture is used, for example, chloroform/methanol in the ratio 2:1 (v/v) for 20 min using a shaker and followed by the addition of mixture, i.e., chloroform/water in the ratio of 1:1 (v/v) for 10 min. Similarly, Pratoomyot et al extracted oil from different algal species using the solvent system chloroform/methanol in the ratio of 2:1 (v/v) and found different fatty acid content. Ryckebosch et al. optimized an analytical procedure and found chloroform/methanol in the ratio 1:1 as the best solvent mixture for the extraction of total lipids. Similarly, Lee et al. extracted lipid from the green alga Botryococcus braunii using different solvent system and obtained the maximum lipid content with chloroform/methanol in the ratio of 2:1. Hossain et al., used hexane/ether in the ratio 1:1 (v/v) for oil extraction and allowed to settle for 24 h. Using a two-step process, reported about 80% of lipid recovery using ethanol and hexane in the two steps for the extraction and purification of lipids. Therefore, a selection of a most suitable solvent system is required for the maximum extraction of oil for an economically viable process.

Lee compared the performance of various disruption methods, including autoclaving, bead-beating, microwaves, sonication, and using 10% NaCl solution in the extraction of Botryococcus sp., Chlorella vulgaris, and Scenedesmus sp, using a mixture of chloroform and methanol (1:1).

Transesterification

This is a process to convert algal oil to biodiesel, which involves multiple steps of reactions between triglycerides or fatty acids and alcohol. Different alcohols such as ethanol, butanol, methanol, propanol, and amyl alcohol can be used for this reaction. However, ethanol and methanol are used frequently for the commercial development due to its low cost and its physical and chemical advantages. The reaction can be performed in the presence of an inorganic catalyst (acids and alkalies) or lipase enzyme. In this method, about 3 mol of alcohol are required for each mole of triglyceride to produce 3 mol of methyl esters (biodiesel) and 1 mol of glycerol (by-product). Glycerol is denser than biodiesel and can be periodically or continuously removed from the reactor in order to drive the equilibrium reaction. The presence of methanol, the co-solvent that keeps glycerol and soap suspended in the oil, is known to cause engine failure. Thus, the biodiesel is recovered by repeated washing with water to remove glycerol and methanol.

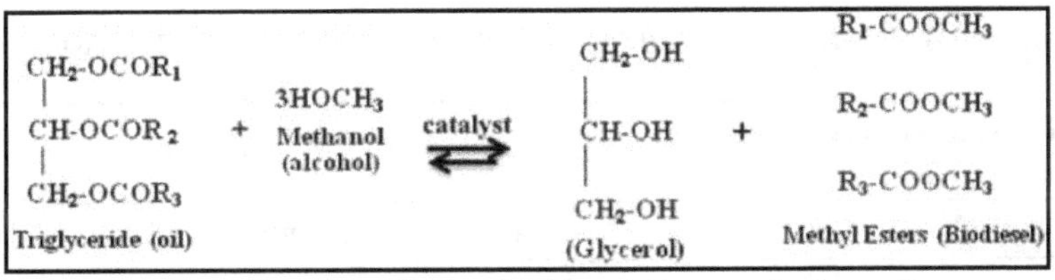

Figure: Transesterification of oil to biodiesel. R_{1-3} are hydrocarbon groups.

The reaction rate is very slow by using the acid catalysts for the conversion of triglycerides to methyl esters, whereas the alkali-catalyzed transesterification reaction has been reported to be 4000 times faster than the acid-catalyzed reaction. Sodium and potassium hydroxides are the two commercial alkali catalysts used at a concentration of about 1% of oil. However, sodium methoxide has become the better catalyst rather than sodium hydroxide.

used Scenedesmus sp. for the biodiesel production through acid and alkali transesterification process. They reported 55.07 ± 2.18%, based on lipid by wt of biodiesel conversion using NaOH as an alkaline catalyst than using H_2SO_4 as 48.41 ± 0.21% of biodiesel production. In comparison to acid and alkalies, lipases as biocatalyst have different advantages as the catalysts due to its versatility, substrate selectivity, regioselectivity, enantioselectivity, and high catalytic activity at ambient temperature and pressure. It is not possible by some lipases to hydrolyze ester bonds at secondary positions, while some other group of enzymes hydrolyzes both primary and secondary esters. Another group of lipases exhibits fatty acids selectivity, and allow to cleave ester bonds at particular type of fatty acids. cloned the lipase gene lipB68 and expressed in Escherichia coli BL21 and further used it as a catalyst for biodiesel production. LipB68 could catalyze the transesterification reaction and produce biodiesel with a yield of 92% after 12 h, at a temperature of 20°C. The activity of the lipase enzyme with such a low temperature could provide substantial savings in energy consumption. However, it is rarely used due to its high cost.

Extractive transesterification. It involves several steps to produce biodiesel such as drying, cell disruption, oils extraction, transesterification, and biodiesel refining. The main problems are related with the high water content of the biomass (over 80%), which overall increases the cost of whole process.

In situ transesterification. This method skips the oil extraction step. The alcohol acts as an extraction solvent and an esterification reagent as well, which enhances the porosity of the cell membrane. Yields found are higher than via the conventional route, and waste is also reduced. Industrial biodiesel production involves release of extraction solvent, which contributes to the production of atmospheric smog and to global warming. Thus, simplification of the esterification processes can reduce the disadvantages of this attractive bio-based fuel. The single-step methods can be attractive solutions to reduce chemical and energy consumption in the overall biodiesel production process. A comparison of direct and extractive transesterification is given in table below.

Table: Comparison of extractive transesterification and *in situ* methods.

Sl. no.	Extractive transesterification	In situ transesterification
1	Low heating value	Heating value is high
2	Product yield is low	Higher product yield
3	Process is complex and time taking	Quick and simple operation process
4	Lipid loss during process	Avoided potential lipid loss
5	Waste water pollutes the environment	Reduced waste water pollutants
6	Production cost is high	Absence of harvesting and dewatering lowers the cost

Bioethanol Production

Several researchers have been reported bioethanol production from certain species of algae, which produce high levels of carbohydrates as reserve polymers. Owing to the presence of low lignin and hemicelluloses content in algae in comparison to lignocellulosic biomass, the algal biomass have been considered more suitable for the bioethanol production. Recently, attempts have been made (for the bioethanol production) through the fermentation process using algae as the feedstocks to make it as an alternative to conventional crops such as corn and soyabean. A comparative study of algal biomass and terrestrial plants for the production of bioethanol has been given in table. There are different micro and macroalgae such as Chlorococcum sp., Prymnesium parvum, Gelidium amansii, Gracilaria sp., Laminaria sp., Sargassum sp., and Spirogyra sp., which have been used for the bioethanol production. These algae usually require light, nutrients, and carbon dioxide, to produce high levels of polysaccharides such as starch and cellulose. These polysaccharides can be extracted to fermentable sugars through hydrolysis and further fermentation to bioethanol and separated through distillation as shown in figure below.

Table: Comparative study between algal biomass and terrestrial plants for bioethanol production.

Feedstock	Conditions	Bioethanol
ALGAE		
Chlorococcum infusionum	Alkaline pre-treatment, temp. 120°C, S. cerevisiae	260 g ethanol/kg algae
Spirogyra	Alkaline pre-treatment, synthetic media growth, saccharification of biomass by Aspergillus niger, fermentation by S. cerevisiae	80 g ethanol/kg algae
Chlorococcum humicola	Acid pre-treatment, temp. 160°C, S. cerevisiae	520 g ethanol/kg microalgae
TERRESTRIAL PLANTS		
Madhuca latifolia	Strain Zymomonas mobilis MTCC 92, immobilized in Luffa cylindrical sponge disks, temp. 30°C	251.1 ± 0.012 g ethanol/kg flowers
Manihot esculenta	Enzyme termamyl and amyloglucosidase, 1 N HCl, Saccharomyces cerevisiae, ca-alginate immobilization	189 ± 3.1 g ethanol/kg flour cassava
Sugarcane bagasse	Acid (H_2SO_4) hydrolysis, Kluyveromyces sp. IIPE453, Fermentation at 50°C	165 g ethanol/kg bagasse
Rice straw	Cellulase, β-glucosidase, solid state fermentation, strain Trichoderma reesei RUT C30, and Aspergillus niger MTCC 7956	93 g ethanol/kg pretreated rice straw

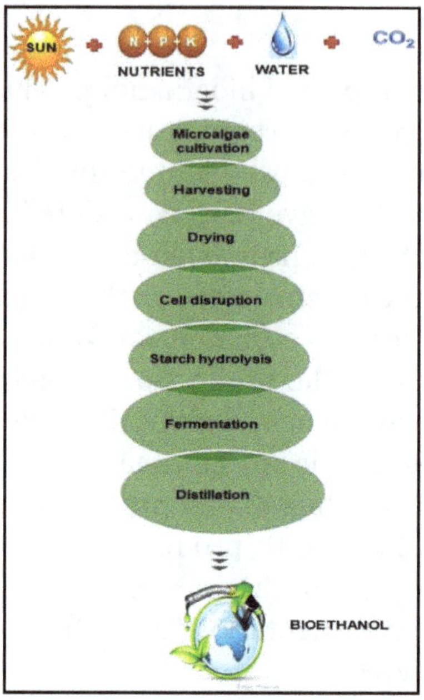

Process for bioethanol production from microalgae.

Pre-treatment and Saccharification

It has been reported that, the cell wall of some species of green algae like Spirogyra and Chlorococcumcontain high level of polysaccharides. Microalgae such as C. vulgaris contains about 37% of starch on dry weight basis, which is the best source for bioethanol with 65% conversion efficiency. Such polysaccharide based biomass requires additional processing like pre-treatment and saccharification before fermentation. Saccharification and fermentation can also be carried out simultaneously using an amylase enzyme producing strain for the production of ethanol in a single step. Bioethanol from microalgae can be produced through the process, which is similar to the first generation technologies involving corn based feedstocks. However, there is limited literature available on the fermentation process of microalgae biomass for the production of bioethanol.

The pre-treatment is an important process, which facilitates accessibility of biomass to enzymes to release the monosaccharides. Acid pre-treatment is widely used for the conversion of polymers present in the cell wall to simple forms. The energy consumption in the pre-treatment is very low and also it is an efficient process found 7% (w/w) H_2SO_4 as the promising concentration for the pre-treatment of the brown macroalgae Nizimuddinia zanardini to obtain high yield of sugars without formation of any inhibitors. Candra and studied the bioethanol production using red seaweed Eucheuma cottonii through acid hydrolysis. In this study, 5% H_2SO_4 concentration was used for 2 h at 100°C, which yielded 15.8 g/L of sugars. However, there are other alternatives to chemical hydrolysis such as enzymatic digestion and gamma radiation to make it more sustainable.

Similar to starch, there are certain polymers such as alginate, mannitol, and fucoidan present in the cell wall of various algae, which requires additional processing like pre-treatment and saccharification before fermentation. Another form of storage carbohydrate found in various brown seaweeds and microalgae is laminarin, which can be hydrolyzed by β-1,3-glucanases or laminarinases. Laminarinases can be categorized into two groups such as exo- and endo-glucanases based on the mode of hydrolysis, which usually produces glucose and smaller oligosaccharides as the end product. Both the enzymes are necessary for the complete digestion of laminarin polymer.

Markou et saccharified the biomass of Spirulina (Arthrospira platensis), fermented the hydrolyzate and obtained the maximum ethanol yield of 16.32 and 16.27% ($g_{ethanol}/g_{biomass}$) produced after pre-treatment with 0.5 N HNO_3 and H_2SO_4, respectively. Yanagisawa et al. investigated the content of polysaccharide materials present in three types of seaweeds such as sea lettuce (Ulva pertusa), chigaiso (Alaria crassifolia), and agar weed (Gelidium elegans). These seaweeds contain no lignin, which is a positive signal for the hydrolysis of polysaccharides without any pre-treatment. Singh and used Spirogyra biomass for the production of bioethanol using Saccharomyces cerevisiae and Zymomonas mobilis. In a method, they followed acid pre-treatment of algal biomass and further saccharified using α-amylase producing Aspergillus niger. In another method, they directly saccharified the biomass without any pre-treatment. The direct saccharification process resulted in 2% (w/w) more alcohol in comparison to pretreated and saccharified algal biomass. This study revealed that the pre-treatment with different chemicals are not required in case of Spyrogyra, which reflects its economic importance for the production of ethanol. Also, cellulase enzyme has been used for the saccharification of algal biomass containing cellulose. However, this enzyme system is more expensive than amylases and glucoamylases, and doses required for effective cellulose saccharification are usually very high. applied different cellulases on green alga Ulva for saccharification and found highest conversion efficiency of biomass into reducing sugars by using cellulase 22119 rather than viscozyme L, cellulase 22086 and 22128.

Fermentation

There are different groups of microorganisms like yeast, bacteria, and fungi, which can be used for the fermentation of the pretreated and saccharified algal biomass under anaerobic process for the production of bioethanol. Nowadays, S. cerevisiae and Z. mobilis have been considered as the bioethanol fermenting microorganisms. However, fermentation of mannitol, a polymer present in certain algae is not possible in anaerobic condition using these well known microorganisms and requires supply of oxygen during fermentation, which is possible only by Zymobacter palmae.

Certain marine red algae contain agar, a polymer of galactose and galactopyranose, which can be used for the production of bioethanol. The biomass of red algae can be depolymerized into different monomeric sugars like glucose and galactose. In addition to mannitol and glucose, brown seaweeds contain about 14% of extra carbohydrates in the form of alginate. Horn et al. reported the presence of alginate, laminaran, mannitol, fucoidan, and cellulose in some brown seaweeds, which are good source of sugars. They fermented brown seaweed extract having mannitol using bacteria Z. palmae and obtained an ethanol yield of about 0.38 g ethanol/g mannitol.

In the literature, there are many advantages supporting microalgae as the promising substrate for bioethanol production. Hon-Nami used Chlamydomonas perigranulata algal culture and obtained different by-products such as ethanol and butanediol. Similarly, Yanagisawa et al. obtained glucose and galactose through the saccharification of agar weed (red seaweed) containing glucan and galactan and obtained 5.5% of ethanol concentration through fermentation using S. cerevisiae IAM 4178. Harun et al. obtained 60% more ethanol in case of lipid extracted microalgal biomass rather than intact algal biomass of Chlorococcum sp. This shows the importance of algal biomass for the production of both biodiesel and bioethanol.

Biogas Production

Recently, biogas production from algae through anaerobic digestion has received a remarkable attention due to the presence of high polysaccharides (agar, alginate, carrageenan, laminaran, and mannitol) with zero lignin and low cellulose content. Mostly, seaweeds are considered as the excellent feedstock for the production of biogas. Several workers have demonstrated the fermentation of various species of algae like Scenedesmus, Spirulina, Euglena, and Ulva for biogas production. The production of biogas using algal biomass in comparison to some terrestrial plants is shown in table below.

Table: Comparative study between algal biomass and terrestrial plants for biogas production.

Feedstock	**Conditions**	**Biogas**
ALGAE		
Blue algae	pH-6.8, microcystin (MC) biodegradation	189.89 mL/g of VS
Chlamydomonas reinhardtii	Drying as the pre-treatment, batch fermentation, temp. 38°C	587 mL/g of VS
Scenedesmus obliquus		287 mL/g of VS
Ulva sp.	Batch reactor, Co-digestion with bovine slurry, temp. 35°C	191 mL/g of VS
Laminaria digitata		246 mL/g of VS
Saccorhiza polyschides		255 mL/g of VS

Feedstock	Conditions	Biogas
Saccharina latissima		235 mL/g of VS
TERRESTRIAL PLANTS		
Banana stem	Pre-treatment: 6% NaOH in 55°C for 54 h. 37 ± 1°C for 40 days, batch	357.9 mL/g of VS
Saline creeping wild ryegrass	35°C for 33 days, batch	251 mL/g of VS
Rice straw	Pre-treatment: ammonia conc. 4% and moisture content 70%, temp. 35 ± 2°C, 65 days,120 rpm, batch	341.35 mL/g of VS
Date palm tree wastes	Pre-treatment: alkaline, particle size 2–5 mm, temp. 40°C	342.2 mL/g of VS

Biogas is produced through the anaerobic transformation of organic matter present in the biodegradable feedstock such as marine algae, which releases certain gases like methane, carbon dioxide, and traces of hydrogen sulfide. The anaerobic conversion process involves basically four main steps. In the first step, the insoluble organic material and higher molecular mass compounds such as lipids, carbohydrates, and proteins are hydrolyzed into soluble organic material with the help of enzyme released by some obligate anaerobes such as Clostridia and Streptococci. The second step is called as acidogenesis, which releases volatile fatty acids (VFAs) and alcohols through the conversion of soluble organics with the involvement of enzymes secreted by the acidogenic bacteria. Further, these VFAs and alcohols are converted into acetic acid and hydrogen using acetogenic bacteria through the process of acetogenesis, which finally metabolize to methane and carbon dioxide by the methanogens.

Sangeetha et al. reported the anaerobic digestion of green alga Chaetomorpha litorea with generation of 80.5 L of biogas/kg of dry biomass under 299 psi pressure. Vergara-Fernandez et al. evaluated digestion of the marine algae Macrocystis pyrifera and Durvillaea antarctica marine algae in a two-phase anaerobic digestion system and reported similar biogas productions of 180.4 (±1.5) mL/g dry algae/day with a methane concentration around 65%. However, in case of algae blend, same methane content was observed with low biogas yield. Mussgnug et al. reported biogas production from some selected green algal species like Chlamydomonas reinhardtii and Scenedesmus obliquus and obtained 587 and 287 mL biogas/g of volatile solids, respectively. Further, there are few studies, which have been conducted with microalgae showing the effect of different pre-treatment like thermal, ultrasound, and microwave for the high production of biogas.

However, there are different factors, which limit the biogas production such as requirement of larger land area, infrastructure, and heat for the digesters. The proteins present in algal cells increases the ammonium production resulting in low carbon to nitrogen ratio, which affects biogas production through the inhibition of growth of anaerobic microorganisms. Also, anaerobic microorganisms are inhibited by the sodium ions. Therefore, it is recommended to use the salt tolerating microorganisms for the anaerobic digestion of algal biomass.

Biohydrogen Production

Recently, algal biohydrogen production has been considered to be a common commodity to be used as the gaseous fuels or electricity generation. Biohydrogen can be produced through different processes like biophotolysis and photo fermentation. Biohydrogen production using algal biomass is comparative to that of terrestrial plants. Park et al. found Gelidium amansii (red alga) as the potential source of biomass for the production of biohydrogen through anaerobic fermentation. Nevertheless, they found 53.5 mL of H_2 from 1 g of dry algae with a hydrogen production rate of 0.518 L H_2/g VSS/day. The authors found an inhibitor, namely, 5-hydroxymethylfurfural (HMF) produced through the acid hydrolysis of G. amansii that decreases about 50% of hydrogen production due to the inhibition. Thus, optimization of the pre-treatment method is an important step to maximize biohydrogen production, which will be useful for the future direction. Saleem et al. reduced the lag time for hydrogen production using microalgae Chlamydomonas reinhardtii by the use of optical fiber as an internal light source. In this study, the maximum rate of hydrogen production in the presence of exogenic glucose and optical fiber was reported to be 6 mL/L culture/h, which is higher than other reported values.

Table: Comparative study between algal biomass and terrestrial plants for biohydrogen production.

Feedstock	Conditions	Biohydrogen
ALGAE		
Gelidium amansii	Hydrolysis at 150°C	53.5 mL of H_2/g of dry algae
Laminaria japonica	Mesophilic condition (35 ± 1°C), pH of 7.5, anaerobic sequencing batch reactor, hydraulic retention time (HRT) of 6 days	71.4 mL H_2/g of dry algae
TERRESTRIAL PLANTS		
Bagasse	Strain Klebsiella oxytoca HP1, temp. 37.5°C, pH-7	107.8 ± 7.5 mL H_2/g bagasse
Corn stalk	Temp. 55°C, pH-7.4	61.4 mL/g of cornstalk
Pretreated wheat straw	Strain Caldicellulosiruptor saccharolyticus, Temp. 70°C, pH-7.2	44.7 mL/g of dry wheat straw
Wheat straw	Acid pre-treatment, simultaneous saccharification and fermentation (SSF)	141 mL/g VS

Some of microalgae like blue green algae have glycogen instead of starch in their cells. This is an exception, which involves oxidation of ferrodoxin by the hydrogenase enzyme activity for the production of hydrogen in anaerobic condition. However, another function of this enzyme is to be involved in the detachment of electrons. Therefore, different researchers have focused for the identification of these enzyme activities having interactions with ferrodoxin and the other metabolic functions for microalgal photo-biohydrogen production. They are also involved with the change of these interactions genetically to enhance the biohydrogen production.

Bio-Oil and Syngas Production

Bio-oil is formed in the liquid phase from algal biomass in anaerobic condition at high temperature. The composition of bio-oil varies according to different feedstocks and processing conditions, which is called as pyrolysis. There are several parameters such as water, ash content, biomass composition, pyrolysis temperature, and vapor residence time, which affect the bio-oil productivity. However, due to the presence of water, oxygen content, unsaturated and phenolic moieties, crude bio-oil cannot be used as fuel. Therefore, certain treatments are required to improve its quality. Bio-oils can be processed for power generation with the help of external combustion through steam and organic rankine cycles, and stirling engines. However, power can also be generated through internal combustion using diesel and gas-turbine engines. In literature, there are limited studies on algae pyrolysis compared to lignocellulosic biomass. Although, high yields of bio-oil occur through fluidized-bed fast pyrolysis processes, there are several other pyrolysis modes, which have been introduced to overcome their inherent disadvantages of a high level of carrier gas flow and excessive energy inputs investigated suitability of the microalgal biomass for bio-oil production and found the superior quality than the wood. produced bio-oil from pyrolysis of algae (Nannochloropsis sp.) at 300°C after lipid extraction, which composed of 50 wt% acetone, 30 wt% methyl ethyl ketone, and 19 wt% aromatics such as pyrazine and pyrrole. Similarly, Choi et al. carried out pyrolysis study on a species of brown algae Saccharina japonica at a temperature of 450°C and obtained about 47% of bio-oil yield.

Gasification is usually performed at high temperatures (800–1000°C), which converts biomass into the combustible gas mixture through partial oxidation process, called syngas or producer gas. Syngas is a mixture of different gases like CO, CO_2, CH_4, H_2, and N_2, which can also be produced through normal gasification of woody biomass. In this process, biomass reacts with oxygen and water (steam) to generate syngas. It is a low calorific gas, which can be utilized in the gas turbines or used directly as fuel. Different variety of biomass feedstocks can be utilized for the production of energy through the gasification process.

Camelina

Camelina [Camelina sativa (L.) Crantz], also known as false flax or gold-of-pleasure, is a broadleaf oilseed flowering plant of the Brassicaceae (mustard) family that grows optimally in temperate climates. Other more common members of this family include important food crops such as broccoli, Brussels sprouts, cabbage, cauliflower, collards, kales, kohlrabi, radish, rapeseed/canola, rutabaga, turnip and various mustards. Camelina can be grown in a variety of climatic and soil conditions as a spring or summer annual or as a biannual winter crop.

Camelina Oil and Meal

Camelina yields anywhere from 336 to 2240 kg of seeds per hectare at maturity with the lipid content of individual seeds ranging between 35 and 45 weight percent (wt%). The resulting yield of camelina oil is thus calculated to be between 106 and 907 liters per hectare, which as seen in table below at the higher end of the range. It is higher than soybean and sunflower oils but lower than rapeseed oil. The wide variability in yield is attributable to such factors as the climate and conditions at the growing location as well as by the amount of agricultural inputs applied to the crop and the presence of pests, weeds and disease. With further genetic modification, the yield of camelina may approach that of rapeseed. Commercially, Great Plains Oil & Exploration, LLC and Sustainable Oils, LLC are the two primary commercial suppliers of camelina oil in the United States.

Table: Comparison of yields from several oilseeds.

	Camelina	Rapeseed	Soybean	Sunflower
Seed yield (T/ha)	0.90–2.24	2.68–3.39a	2.14–2.84a	1.44–1.70a
Oil content (wt%)	35–45	40–44	18–22	39–49
Oil yield (L/ha)b	106–907	965–1342	347–562	505–750

a. Oil World Annual 2009, 1, Oilseeds 5–9 (T = metric ton = 1000 kg). Data from EU-27 from 2002–2006.

b. Calculated by the author (BRM) from yield and oil content data.

α Linolenic acid generally comprises between 32 and 40 wt% of the fatty acid composition of camelina oil (Table). Other fatty acids in quantities above 10 wt% include linoleic, oleic, and 11-eicosenoic acids. The acid value of crude camelina oil extracted either by solvent (hexane) or cold press is typically between 1 and 5 mg KOH/g of oil, which is considerably lower than many other crude plant oils. Camelina oil has been used for numerous applications including directly as an experimental fuel for diesel transport engines, as a culinary oil, and as a biological feedstock for the experimental production of longchain wax esters for potential cosmetic and lubricant applications. Camelina oil may also be used as an industrial source of a-linolenic acid, as it has a comparatively high content of this constituent versus the commodity oils.

Extraction of oil from camelina seeds by mechanical expeller yields a meal that consists of approximately 10% residual oil, 45% crude protein, 13% fibers, 5% minerals, and other minor constituents such as glucosinolates and vitamins. The concentration of glucosinolates in dry camelina seeds ranges from 13 to 36 lmol/g. When ingested in sufficient quantity, glucosinolates cause deleterious effects in animals such as reduced palatability as well as decreased growth and production. Consequently, camelina meal cannot exceed 10 wt% of the total food ration given to feedlot beef cattle and broiler chickens in the United States. However, even a 10% ceiling represents significant market potential where large numbers of animals are raised for human consumption, thus representing an important revenue stream for camelina meal.

Table: Fatty acid composition (%) of camelina oil from several literature

Fatty acid	Carbon number	Leonard	Moser & Vaughn	Zuhr & Matthaus	Frohlich & Rice
Palmitic	16:0	5.3	6.8	5.4	5.4
Stearic	18:0	2.5	2.7	2.5	2.6
Oleic	18:1	12.6	18.6	14.9	14.3
Linoleic	18:2	15.6	19.6	15.2	14.3
α-Linolenic	18:3	37.5	32.6	36.8	38.4
Arachidic	20:0	1.2	1.5	1.3	1.4
11-Eicosenoic	20:1	15.5	12.4	15.5	16.8
Eicosadienoic	20:2	2.0	1.3	1.9	
Eicosatrienoic	20:3	1.7	0.8	1.6	
Behenic	22:0	0.3	0.2	0.3	0.2
Erucic	22:1	2.9	2.3	2.8	2.9
Others	–	2.9	1.2	1.8	3.7

Camelina Oil as Feedstock for Biodiesel

Camelina is particularly attractive as an alternative feedstock for biodiesel production as a result of its low cost versus commodity oils coupled with its potential to significantly enhance domestic feedstock availability.

Biodiesel is classically prepared by transesterification of lipids in the presence of an homogenous alkali catalyst and excess methanol at elevated (608C) temperature [3–5]. figure below illustrates a typical schematic for industrial production of biodiesel. Crude feedstocks that have low free fatty acid contents (less than 3.0 wt%) can be directly transesterified without pretreatment, thereby eliminating a costly pretreatment step. Camelina is one such crude oil and has been successfully converted to biodiesel (fatty acid methyl esters) by the classic method as well as with heterogeneous metal oxide catalysts both with and without microwave irradiation and at non-catalytic sub- and supercritical conditions employing co-solvents with methanol. The fuel properties (cold flow properties, oxidative stability, kinematic viscosity, cetane number, etc) of camelina-based biodiesel are similar to those of biodiesel prepared from soybean oil, thus indicating its acceptability for use as biodiesel. Additionally, fatty acid ethyl esters have been prepared from camelina oil and along with methyl esters were evaluated as blend components in petrodiesel (a15 ppm S). As was the case with the neat esters, camelina-based biodiesel blends in petrodiesel exhibited fuel properties comparable to the corresponding soybean-based blends. Commercially, Green Earth Fuels, LLC in partnership with Sustainable Oils, LLC and INEOS in partnership with Great Plains Oil & Exploration, LLC are among the industrial producers of fatty acid methyl esters from camelina oil.

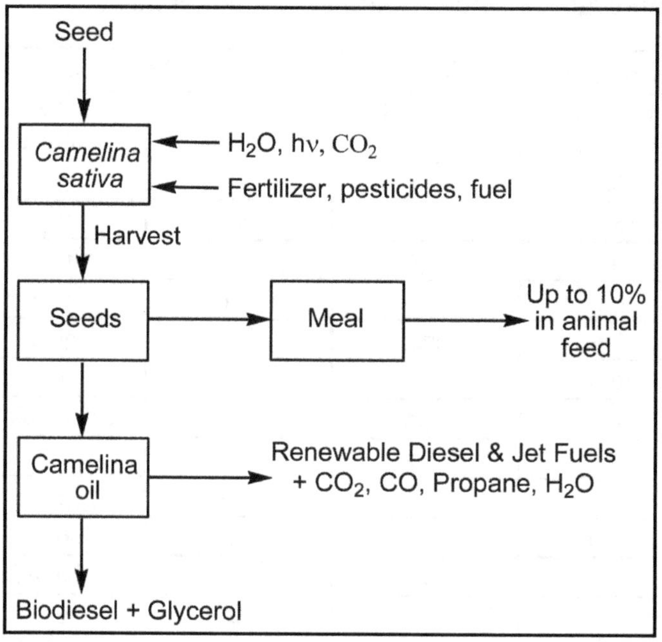

Life cycle diagram for production of alternative diesel fuels from camelina.

Jatropha

Jatropha curcas Linnaeus is a species of flowering plant in the spurge family, Euphorbiaceae; the tree is shown in figure and its seeds are shown in figure. It is a tree or shrub that is native to the American tropics, mostly Mexico and Central America, but it grows under a variety of agroclimatic conditions and is commonly found in most of the tropical and subtropical regions of the world. Thus, it ensures a reasonable production of seeds with very minor care. The oil content of Jatropha seed ranges from 30% to 35% by weight. The common by-products produced while processing the biodiesel are glycerol and oilseed cake.

These by-products shall reduce the cost of biodiesel depending upon the price which these products can fetch. The cost components of biodiesel are the price of seed, seed collection and oil extraction, oil trans-esterification, transport of seed and oil. The cost of biodiesel produced by trans-esterification of oil obtained from Jatropha curcas seeds will be very close to the cost of seed required to produce the quantity of biodiesel as the cost of extraction of oil and its processing in to biodiesel is recoverable to a great extent from the income of oil cake and glycerol which are by-products.

Rapeseed

Rapeseed is related to mustard and to other cabbage-family crops. Rapeseed has been cultivated since the 20th century B.C. Because the plant can grow with less sunlight and at lower temperatures than other crops, it was cultivated in Europe as early as the 13th century A.D.

Rapeseed oil has been used for cooking, lighting, and industrial uses. However, traditional rapeseed contains high quantities of erucic acid and glucosinolates, which make the seed meal unpalatable and possibly dangerous to livestock if fed in large quantities.

The Technological Process of Obtaining Biodiesel

Preparation of Rapeseed Oil

The biodiesel is produced mainly from rapeseed with doubly improved varieties of seeds. The composition of rapeseed oil must meet specific quality requirements that are included in the standard for methyl esters as fuels for diesel engines.

Vegetable oils can be obtained by mechanical or/and chemical extraction. Crushing of seeds is a traditional way of producing oils using different types of presses. However, chemical extraction is a modern way of obtaining vegetable oils using solvent extracts (e.g. hexane). These methods can be combined e.g. residue after pressing is subjected to solvent extraction. Some smaller plants are rather only press-plants without applying the step of solvent extraction.

The process of rapeseed oil preparation consists of seed cleaning, preconditioning, flaking, cooking, screw pressing, solvent extraction, desolventizing, distillation and degumming. Rapeseed cleaning involves aspiration, screen separation to remove oversized particles, and screen separation to remove undersized particles. Next, the whole seeds are preheating prior to processing (to about 30- 40°C) by indirect heating or direct hot air contact. Cell walls must be ruptured in order to extract the oil. Moreover, it allows a solvent to get into cellular structure dissolving and diluting the lipid portions which then are separated from the solid flake. Next, the preheated rapeseed oil is flaked between two smooth surface cast-iron rolls. Extraction of oil from flaked rapeseed proceed mostly by pre-press solvent extraction. Then there is further extraction of oil seeds and press cake with hexane. In the next step, the hexane solvent is removed from the extracted cake and then distilled from the rapeseed oil. The phospholipids or gums need to be removed in a degumming step. Next, free from moisture and cooled oil is transferred to the refining process or into a storage. Crude oil obtained by extraction contain many undesirable substances such as mucoid substances, coloring matters and free fatty acids. Therefore, it has to be refined to obtain a high quality oil. The refining process involves degumming, neutralization, drying, bleaching, and deodorization. Gums compose about 2% of solvent-extracted rapeseed oil. Degumming treatment uses hot water or steam and phosphoric acid, citric acid, or other acidic materials. Precipitated gums are removed by centrifugation. Free fatty acids can be neutralized with alkali solution. After that, the oil is washed with hot water to remove traces of soaps that can reduce stability of oil. Next, the oil is dried to remove traces of water. Bleaching process proceeds by adsorption of the color producing substances on an adsorbent material such as bentonite or Fuller's earth. The next step i.e. deodorization is a vacuum steam distillation process in order to remove trace constituents giving rise to undesirable odors in oils.

The high contents of free fatty acids and water in the collected oil are responsible for secondary reactions during transesterification. Therefore, a pretreatment of oil is necessary before load the reactor to produce biodiesel. Oil refining aims to remove the excess of phospholipids, salts of iron or copper and obtain low peroxide number as well as low acid number (less than 1mg KOH/g) i.e. low content of free fatty acids which react with basic catalyst during transesterification to form soap making difficult purification of glycerin phase and increasing the demand for catalyst. The presence of phospholipids in the oil increases its resistance to oxidation, but also emulsifying reaction system makes it difficult separation of glycerol and ester phase. The phosphorus content in the oil directed to transesterification shouldn't exceed 10 ppm. Rapeseed oil led to methanolysis should be deprived of moisture (water content<0.5%), because its presence causes the hydrolysis of triacylglycerol, resulting in the formation of free fatty acids. Some factors decide about quality of rape seeds i.e. the degree of seeds maturity, the presence of damaged seeds, seed moisture content and storage times and conditions. Higher free fatty acids content can be stated in the seeds of inferior quality.

Transesterification of Rapeseed Oil

Transesterification is the best way to obtain biodiesel because it is well-known and cheap process which gives less problems for the engines than another methods. In a standard process of production biodiesel from rapeseed oil there are following process steps i.e. esterification of rapeseed oil, separation of esterification products, methanol distillation and purification of the ester. The main stage of the process is based on the transesterification reaction of rapeseed oil with an alcohol (methanol, ethanol) which results in formation of esters of alcohols and glycerol. The reaction is reversible due to formation of water, which is responsible for shifting the equilibrium towards the reagents. In order to move the chemical equilibrium towards the ester, an excess of alcohol is used. The transesterification involves three consecutive and reversible reactions. Each molecule of alcohol makes with the residue of fatty acid a distinct monoester molecule. If the triglyceride contains three different fatty acids in its molecule, a mixture of rapeseed oil esters is obtained.

There are used various chemical catalysts in the transesterification reaction i.e. acids, alkalis and enzymes. Some of them are the most effective i.e. alkaline catalysts and their methoxides. Only anhydrous reactants should be used, because the water decomposes catalyst and leads the reaction to the side of the reactants. The amount of the catalyst depends on its type, quality of substrates, reaction time and temperature. Its value varies from 0.2% to 2% by mass relative to the weight of oil. Unrefined oil requires more catalyst. Free fatty acids content in oil shouldn't exceed 0.5-1%. The higher the water content in methanol, the greater the consumption of catalyst and higher contents of free fatty acids as well as soaps. Forming of soaps reduces the activity of the catalyst and increases the viscosity, so the separation of the glycerol is more difficult.

In the majority of the biodiesel production processes, methanol is used, mainly due to the cost and its physical and chemical advantages (polar, shortest chain alcohol and the reactivity is easier). For the basic-catalysed process, the optimal ratio is 6:1 (alcohol: oil). Applying alkaline catalysts in the reaction allow to carry out the transesterification process at room temperature. The reaction time is very diverse and can vary from several minutes to several hours (typically about 30-60 minutes). Mixing is important because of the lack of mutual solubility of the substrates. Appropriately intensive mixing of reactants increases the contact area. Using the refined oils with a higher purity as well as pure and anhydrous methanol, allows to obtain biodiesel with a high content of methyl ester (96.5- 99%).

In the transesterification reaction, the oil is converted to FAME (Fatty Acid Methyl Ester). The flow, which goes out from the reactor, should be purified. First, the methanol is recovered (94%) and then residue is carried to separation of glycerin. Next, the obtained biodiesel is purified to commit the standards of quality.

The primary by-products generated during the production of biodiesel include: oilseed meal, straw and glycerol. Oilseed meal is produced in the process of pressing oil from rapeseeds. It accounts for 70% by weight, and oil represents the remaining 30%. Oilseed meal can be used as a component of feed or as a component of the binder for the production of smokeless fuel, additive to fuel briquettes and as a basic component of mat to cover mine dumps.

Straw is formed at the beginning of the production cycle. Part of straw in the form of chaff can be plowed already on the field and is a valuable organic fertilizer. Properly prepared straw can also be used as a component of the roughage. Straw is also a valuable energy source, because its calorific value is about 30-40% of the calorific value of coal and the ash formed of its combustion can be used as a mineral fertilizer.

Glycerol is a byproduct formed in the esterification of rapeseed oil. In this process also are created soaps and free fatty acids. Glycerol phase constituting concentrated glycerol (about 80% solution) may be transferred to facilities specializing in treatment of glycerol (e.g. cosmetic, pharmaceutical). The cost of obtaining the crude glycerol during esterification of vegetable oils is lower than the cost of traditional methods that had been previously used in splitting of fats. Glycerol from vegetable oils is used both for cosmetic purposes and pharmacology. More often it is an ecological material for the manufacture of certain foods and it is used for making explosives. Using own purification of glycerol, receives an additional by-product in the form of valuable mineral fertilizer. In the case of small factories, where it is uneconomic to build installation of treatment and it is not organized buying glycerol, it can be disposed of by diluting with water in a ratio of 1:100 and added to the slurry.

Soaps and free fatty acids also may be forwarded to the appropriate processing plants, or used in their own farm. Mud can be a valuable addition to the feed for producing Californian earthworms Biohumus – mineral fertilizer.

Vegetable oil is low-density-energy biomass. It limits the scale of biodiesel production. However, there are advantages of biodiesel plant refinery such as low investment and rapid construction.

The main application of fatty acid methyl esters is biodiesel, generally as mixture with petro-diesel for engine use. After refinement, fatty acid methyl esters can be also used as biodegradable solvents. Additionally, there are a source of many commodity chemicals such as higher alcohols. Glycerol is an important chemical product mainly used in areas such as food, beverage, pharmaceutical, cosmetic, tobacco and paper making.

Animal Fat

About one-third of the fats and oils produced in the United States are animal fats. This includes beef tallow, pork lard, and chicken fat. Animal fats are attractive feedstocks for biodiesel because their cost is substantially lower than the cost of vegetable oil. This is partly because the market for animal fat is much more limited than the market for vegetable oil, since much of the animal fat produced in the U.S. is not considered edible by humans.

Animal fat is currently added to pet food and animal feed, and used for industrial purposes such as soap making. Much of the domestic animal fat supply is exported.

Animal fat feedstocks can be made into high-quality biodiesel that meets the ASTM specificationsfor biodiesel. However, there are some drawbacks and challenges to using animal fat feedstocks.

Processing of Animal Fats

Waste fat from animal carcasses are removed and then made into an oil using a rendering process. Rendering consists of grinding the animal by-products to a fine consistency and cooking them until the liquid fat separates and pathogens are destroyed. The solids are usually passed through a screw press to complete the removal of the fat from the solid residue. The cooking process also removes water, which makes the fat and solid material stable against rancidity. The end products are fat, and a high-protein feed additive known as "meat and bone meal."

Fatty Acid Content of Animal Fats

Animal fats are highly saturated, which means that the fat solidifies at a relatively high temperature. Therefore, biodiesel made from animal fat has a high cloud point. For example, biodiesel made from beef tallow and pork lard has a cloud point in the range of 55°F to 60°F. B100 (pure biodiesel) made from animal fat should only be used in a very warm climate. However, animal fat biodiesel can be blended with petro-diesel. At lower blends such as B5 (a blend of 5% biodiesel with 95% petro-diesel), the high cloud point of the animal fat biodiesel does not have much effect on the cloud point of the blend.

The composition, or fatty acid profile, of various animal fats is shown in table below. These data show how saturated the fat will be, which in turn determines how readily it will solidify as the temperature decreases.

Table: Fatty Acid Percentages in Animal Fats.

Fatty acid	*Beef tallow*	*Pork lard*	*Chicken fat*
Myristic 14:0	1.4 - 6.3	0.5 – 2.5	1
Palmitic 16:0	20 - 37	20 – 32	25
Palmitoleic 16:1	0.7 – 8.8	1.7 – 5	8
Stearic 18:0	6 – 40	5 – 24	6
Oleic 18:1	26 – 50	35 – 62	41
Linoleic 18:2	0.5 - 5	3 – 16	18

Beef tallow and pork lard are typically about 40% saturated (sum of myristic, palmitic and stearic acids). Chicken fat is lower at about 30-33%. For comparison, soybean oil is about 14% saturated and canola oil is only 6%. Thus, tallow and lard are usually solid at room temperature and chicken fat, while usually still liquid, is very viscous and nearly solid.

When the animal fat is made into biodiesel, the concern about solidification at lower temperatures continues. The methyl esters from the saturated fatty acids, mainly methyl stearate and methyl palmitate, have high melting points.

Oxidative Stability of Animal Fat Biodiesel

In theory, the saturated fatty acids in animal fats should contribute to better oxidative stability for biodiesel. Animal fats contain very little of the polyunsaturated fatty acids, such as linoleic acid and linolenic acid, that make vegetable oils such as soybean oil and linseed oil so prone to rancidity.

However, in practice animal fat is not always more stable than vegetable oil, because vegetable oils often contain natural anti-oxidants. For example, a test of the peroxide content of lard and vegetable oil showed that the lard experienced oxidation faster than the vegetable oil.

Animal fats contain very little of the natural anti-oxidants such as vitamin E that protect vegetable oils. In some cases, used cooking oils may also contain artificial antioxidants that are added to oil to extend their life, and these artificial anti-oxidants can in turn extend the life of biodiesel made from used cooking oil.

High Cetane Number of Animal Fat Biodiesel

Animal fat feedstocks result in biodiesel with a high cetane number, which is an important quality parameter for diesel fuels. The saturated fatty acids are the source of this high cetane number and values over 60 are common. Soybean oil based biodiesel usually has a cetane number of about 48-52 and petroleum-based diesel fuel is usually between 40 and 44. When animal fat biodiesel is blended with petro-diesel, this high cetane number can help the engine start more quickly and run more quietly.

Lower Nitrous Oxide Emissions from Animal Fat Biodiesel

One of the important attributes of biodiesel is that it lowers the levels of harmful pollutants in the exhaust of diesel engines. One exception to this is oxides of nitrogen (NOx), which are implicated in ozone and smog formation. Biodiesel tends to emit slightly more nitrous oxides than petro-diesel. Many reasons have been cited for this increase in NOx and there is still considerable debate.

However, it has been shown that biodiesel from animal fats tends to produce a smaller increase in NOx and in some cases no increase. The primary reason for this is probably that animal fat biodiesel has a high cetane number (>60) compared with vegetable oil biodiesel. Higher cetane number is known to lower NOx by lowering temperatures during the critical early part of the combustion process.

Contaminants in Animal Fat Feedstocks

Animal fats can be challenging feedstocks for biodiesel because they frequently contain contaminants that should be removed before the fuel is used in an engine.

Phospholipids, or gums, will cause insoluble precipitates when they come into contact with water. Since these precipitates will plug fuel filters, they must be removed from the fuel. Most of this material will separate with the glycerin during processing, or be removed during the purification step (water washing or ion exchange), so it is rare for these compounds to be found in the fuel. In fact, since phospholipids can de-activate exhaust aftertreatment devices on diesel vehicles, they are usually removed from the feedstock before it is converted to fuel. Most commonly, the gums are removed by adding water and citric or phosphoric acid to the feedstock and then separating the precipitates with a centrifuge.

Polymers that are formed naturally at the high temperatures of the rendering process can contribute to a higher viscosity in biodiesel made from animal fat. This higher viscosity can sometimes prevent animal fat biodiesel from passing the viscosity ASTM specification. If this is the case, the animal fat biodiesel can be blended with a lower-viscosity biodiesel.

An occasional problem with rendered fats is polyethylene in the fat which originates from plastic bags, ear tags, or other plastic that is mixed with the animal byproducts. Finely divided polyethylene has been found to cause cloudiness in the fuel and could plug the fuel filter. Producers who are used to working with vegetable oil may be confused by this, since with vegetable oil biodiesel, cloudiness is caused mainly by water in the fuel. If animal fat biodiesel shows cloudiness even after water is removed, this may be caused by stray polyethylene, which can be removed by passing the biodiesel through a fine filter.

High Sulfur Content of Animal Fat Biodiesel

Sulfur can sometimes be a problem for animal fat-based biodiesel. Biodiesel sold for on-highway use is only allowed to contain up to 15 ppm of sulfur. Some samples of beef tallow have been found to contain over 100 ppm of sulfur and chicken fat frequently contains a similar amount. The sulfur apparently originates from sulfur-containing amino acids associated with proteins that carry over from the rendering process.

Measurements of the sulfur levels in biodiesel produced from animal fats have shown that the sulfur level usually decreases by about half when the conversion takes place. However, the remaining sulfur can be difficult to remove. Vacuum distillation is about the only reliable technique for removing this sulfur.

Alternatively, biodiesel high in sulfur can be sold for non-highway use, such as fuel for a boiler or heater.

Paper Waste

Recycled paper, paper sludge from production and even sawdust from the early processing stages of paper have all been examined as possible sources of biofuel. While producers would have to compete with paper companies -- who use the waste in their own recycling programs --researchers believe that some waste could be diverted for fuel production. However, many of these materials might be more trouble than they are worth. that paper is difficult to convert to liquid fuel because of its waxy coating and how the paper is made. "Have you ever tried to light up a newspaper?"."It will burn but then goes out right away because of the form of the paper. It's a complex system; it's a lot harder [to burn] than wood or corn stalks." Paper waste rounds out the bottom of the countdown because it's a source that is already being widely used for products other than biofuel. Combined with the fact that it is difficult to process and somewhat expensive makes this biofuel a limited opportunity.

Materials and Methods

Collection of Substrate

Newspaper, whichwas used as a substrate for the production of bioethanol,was collected from the households.The substrate was collected in a dust free and fungus-free state and was dried in sunlight and was made into small pieces and stored in sealed plastic bags.

Chemical Analysis of the Substrate

Composition of the substrate and its properties were analyzed before pre-treatment. The cellulose content and total carbohydrate in the substrate was estimated by anthrone method, Moisture content and ash content of the substrate were also estimated using standard methods.

Optimization of Pretreatment Process

The pre-treatment optimization for the substrate was carried out by using different combination dilute sulphuric acid ranging from 0 to 6% and heating period of 30, 45 and 60 minutes at 1210 C and 15lb pressure. 1gm of substrate was added with 10 ml of dilute sulphuric acid (1:10). Cellulose released during this optimization process was analysed by anthrone method. After the release of maximum amount of cellulose during pre-treatment process, the solution was taken for hydrolysis.

Hydrolysis of the Pretreated Substrate

Maximum cellulose released during the pretreatment was hydrolysed by the isolated cellulose degrading bacteria. The pretreated substrate was washed with distilled water several times to neutralise the acid concentration. The substrate was oven dried till constant weight and the pH was adjusted to 7.0. Comparison study between isolated cellulose degrading bacteria and the pure culture, CH was performed. A 24hr grown inoculum of isolated cellulose degrading bacteria and pure culture, CH were added to the pretreated substrate. Reducing sugars release during substrate hydrolysis were analysed by Dinitrosalicylic Acid (DNS) method every 24hr from zero hour, for both the organisms. Maximum sugars released during this period were further taken for fermentation to produce bioethanol.

Fermentation of Hydrolysed Broth

Fermentation was carried out using commercially available yeast, Saccharomyces cerevisiae. The pH of hydrolysed broth was adjusted to 4.6 and an inoculum of active yeast (in log phase) was added to the hydrolysed broth. The fermentation was carried out at 360 C until maximum sugars are converted into bioethanol. The reducing sugar utilization during fermentation was analysed by DNS method and the bioethanol production was analysed by using specific gravity method.

Calculation for specific gravity:

$$\frac{W2 - W1 =}{W3 - W1} \textit{Specific Gravity}$$

Where,

W1 = empty weight of specific gravity bottle.

W2 = Weight of sample + specific gravity bottle.

W3 = Weight of distilled water + specific gravity bottle.

Also the ethanol yield was estimated using High Performance Liquid Chromatography (HPLC). The instrument used was "SHIMADZU-C-10AVP".Ethanol was analyzed by using HPLC with cation exchanger SugarPak column (C18). Acetone-Nitrate and water (80:20) was used as a mobile phase at a flow rate of 1mL/min. The injection volume was 5μl and the column temperature was maintained at 90°C. All the samples were filtered through a 0.45μm filter before subjected to HPLC analysis. The evaluate was detected by a refractive index detector at 50°C.

Results and Discussions

Chemical Analysis of the Substrate

The chemical analysis of the substrate showed that the substrate consisted of 45% cellulose. Ash content and moisture content were found to be 5.17% and 6% respectively.

Optimization of Pretreatment of the Substrate

In pre-treatment with dilute sulphuric acid, the structure of the cellulosic biomass will be altered to make cellulose more accessible to the enzymes that convert the carbohydrate polymers into fermentable sugars rapidly and with greater yield.

References

- Animal-fats-for-biodiesel-production-30256: extension.org, Retrieved 22, August 2020
- Sugarcane-biomass: bioenergyconsult.com, Retrieved 09, April 2020
- Top-10-sources-for-biofuel-1769457447: seeker.com, Retrieved 27, April 2020
- Corn-for-biofuel-production-27536: extension.org, Retrieved 19, February 2020
- Rapeseed-and-canola-for-biodiesel-production-26629: extension.org, Retrieved 16, July 2020

Chapter 5

Diverse Areas of Biomass Application

The applications of biomass dealt within this chapter are biogas, biochar, bioliquids, blue carbon, straw and woodchips. The mixture of gases produced by the breaking of organic matter is known as biogas. Biochar is used as a product that is added to soil to improve it. The various aspects elucidated in this text are of vital importance, and provides a better understanding of biomass.

Biogas

Pipes carrying biogas (foreground), natural gas and condensate.

Biogas typically refers to a mixture of different gases produced by the breakdown of organic matter in the absence of oxygen. Biogas can be produced from raw materials such as agricultural waste, manure, municipal waste, plant material, sewage, green waste or food waste. Biogas is a renewable energy source and in many cases exerts a very small carbon footprint.

Biogas can be produced by anaerobic digestion with anaerobic organisms, which digest material inside a closed system, or fermentation of biodegradable materials.

Biogas is primarily methane (CH_4) and carbon dioxide (CO_2) and may have small amounts of hydrogen sulfide (H_2S), moisture and siloxanes. The gases methane, hydrogen, and carbon monoxide (CO) can be combusted or oxidized with oxygen. This energy release allows biogas to be used as a fuel; it can be used for any heating purpose, such as cooking. It can also be used in a gas engine to convert the energy in the gas into electricity and heat.

Biogas can be compressed, the same way natural gas is compressed to CNG, and used to power motor vehicles. In the UK, for example, biogas is estimated to have the potential to replace around 17% of vehicle fuel. It qualifies for renewable energy subsidies in some parts of the world. Biogas can be cleaned and upgraded to natural gas standards, when it becomes bio-methane. Biogas is considered to be a renewable resource because its production-and-use cycle is continuous, and it generates no net carbon dioxide. Organic material grows, is converted and used and then regrows in a continually repeating cycle. From a carbon perspective, as much carbon dioxide is absorbed from the atmosphere in the growth of the primary bio-resource as is released when the material is ultimately converted to energy.

Production

Biogas production in rural Germany.

Biogas is produced as landfill gas (LFG), which is produced by the breakdown of Biodegradable waste inside a landfill due to chemical reactions and microbes, or as digested gas, produced inside an anaerobic digester. A *biogas plant* is the name often given to an anaerobic digester that treats farm wastes or energy crops. It can be produced using anaerobic digesters (air-tight tanks with different configurations). These plants can be fed with energy crops such as maize silage or biodegradable wastes including sewage sludge and food waste. During the process, the microorganisms transform biomass waste into biogas (mainly methane and carbon dioxide) and digestate. The biogas is a renewable energy that can be used for heating, electricity, and many other operations that use a reciprocating internal combustion engine, such as GE Jenbacher or Caterpillar gas engines. Other internal combustion engines such as gas turbines are suitable for the conversion of biogas into both electricity and heat. The digestate is the remaining inorganic matter that was not transformed into biogas. It can be used as an agricultural fertiliser.

There are two key processes: mesophilic and thermophilic digestion which is dependent on temperature. In experimental work at University of Alaska Fairbanks, a 1000-litre digester using psychrophiles harvested from "mud from a frozen lake in Alaska" has produced 200–300 liters of methane per day, about 20%–30% of the output from digesters in warmer climates.

Dangers

The dangers of biogas are mostly similar to those of natural gas, but with an additional risk from the toxicity of its hydrogen sulfide fraction. Biogas can be explosive when mixed in the ratio of one part biogas to 8-20 parts air. Special safety precautions have to be taken for entering an empty biogas digester for maintenance work.

It is important that a biogas system never has negative pressure as this could cause an explosion. Negative gas pressure can occur if too much gas is removed or leaked; Because of this biogas should not be used at pressures below one column inch of water, measured by a pressure gauge.

Frequent smell checks must be performed on a biogas system. If biogas is smelled anywhere windows and doors should be opened immediately. If there is a fire the gas should be shut off at the gate valve of the biogas system.

Landfill Gas

Landfill gas is produced by wet organic waste decomposing under anaerobic conditions in a biogas.

The waste is covered and mechanically compressed by the weight of the material that is deposited above. This material prevents oxygen exposure thus allowing anaerobic microbes to thrive. This gas builds up and is slowly released into the atmosphere if the site has not been engineered to capture the gas. Landfill gas released in an uncontrolled way can be hazardous since it can become explosive when it escapes from the landfill and mixes with oxygen. The lower explosive limit is 5% methane and the upper is 15% methane.

The methane in biogas is 20 times more potent a greenhouse gas than carbon dioxide. Therefore, uncontained landfill gas, which escapes into the atmosphere may significantly contribute to the effects of global warming. In addition, volatile organic compounds (VOCs) in landfill gas contribute to the formation of photochemical smog.

Technical

Biochemical oxygen demand (BOD) is a measure of the amount of oxygen required by aerobic micro-organisms to decompose the organic matter in a sample of water. Knowing the energy density of the material being used in the biodigester as well as the

BOD for the liquid discharge allows for the calculation of the daily energy output from a biodigester.

Another term related to biodigesters is effluent dirtiness, which tells how much organic material there is per unit of biogas source. Typical units for this measure are in mg BOD/litre. As an example, effluent dirtiness can range between 800–1200 mg BOD/litre in Panama.

From 1 kg of decommissioned kitchen bio-waste, 0.45 m^3 of biogas can be obtained. The price for collecting biological waste from households is approximately €70 per ton.

Composition

Typical composition of biogas		
Compound	Formula	%
Methane	CH 4	50–75
Carbon dioxide	CO 2	25–50
Nitrogen	N 2	0–10
Hydrogen	H 2	0–1
Hydrogen sulfide	H 2S	0–3
Oxygen	O 2	0–0.5

The composition of biogas varies depending upon the origin of the anaerobic digestion process. Landfill gas typically has methane concentrations around 50%. Advanced waste treatment technologies can produce biogas with 55%–75% methane, which for reactors with free liquids can be increased to 80%-90% methane using in-situ gas purification techniques. As produced, biogas contains water vapor. The fractional volume of water vapor is a function of biogas temperature; correction of measured gas volume for water vapor content and thermal expansion is easily done via simple mathematics which yields the standardized volume of dry biogas.

In some cases, biogas contains siloxanes. They are formed from the anaerobic decomposition of materials commonly found in soaps and detergents. During combustion of biogas containing siloxanes, silicon is released and can combine with free oxygen or other elements in the combustion gas. Deposits are formed containing mostly silica (SiO 2) or silicates (Si xOy) and can contain calcium, sulfur, zinc, phosphorus. Such *white mineral* deposits accumulate to a surface thickness of several millimeters and must be removed by chemical or mechanical means.

Practical and cost-effective technologies to remove siloxanes and other biogas contaminants are available.

For 1000 kg (wet weight) of input to a typical biodigester, total solids may be 30% of the wet weight while volatile suspended solids may be 90% of the total solids. Protein would be 20% of the volatile solids, carbohydrates would be 70% of the volatile solids, and finally fats would be 10% of the volatile solids.

Benefits of Manure Derived Biogas

High levels of methane are produced when manure is stored under anaerobic conditions. During storage and when manure has been applied to the land, nitrous oxide is also produced as a byproduct of the denitrification process. Nitrous oxide (N2O) is 320 times more aggressive as a greenhouse gas than carbon dioxide and methane 25 times more than carbon dioxide.

By converting cow manure into methane biogas via anaerobic digestion, the millions of cattle in the United States would be able to produce 100 billion kilowatt hours of electricity, enough to power millions of homes across the United States. In fact, one cow can produce enough manure in one day to generate 3 kilowatt hours of electricity; only 2.4 kilowatt hours of electricity are needed to power a single 100-watt light bulb for one day. Furthermore, by converting cattle manure into methane biogas instead of letting it decompose, global warming gases could be reduced by 99 million metric tons or 4%.

Applications

A biogas bus in Linköping, Sweden.

Biogas can be used for electricity production on sewage works, in a CHP gas engine, where the waste heat from the engine is conveniently used for heating the digester; cooking; space heating; water heating; and process heating. If compressed, it can replace compressed natural gas for use in vehicles, where it can fuel an internal combustion engine or fuel cells and is a much more effective displacer of carbon dioxide than the normal use in on-site CHP plants.

Biogas Upgrading

Raw biogas produced from digestion is roughly 60% methane and 29% CO_2 with trace elements of H_2S; it is not of high enough quality to be used as fuel gas for machinery. The corrosive nature of H_2S alone is enough to destroy the internals of a plant.

Methane in biogas can be concentrated via a biogas upgrader to the same standards as fossil natural gas, which itself has to go through a cleaning process, and becomes *biomethane*. If the local gas network allows, the producer of the biogas may use their distribution networks. Gas must be very clean to reach pipeline quality and must be of the correct composition for the distribution network to accept. Carbon dioxide, water, hydrogen sulfide, and particulates must be removed if present.

There are four main methods of upgrading: water washing, pressure swing adsorption, selexol adsorption, and amine gas treating. In addition to these, the use of membrane separation technology for biogas upgrading is increasing, and there are already several plants operating in Europe and USA.

The most prevalent method is water washing where high pressure gas flows into a column where the carbon dioxide and other trace elements are scrubbed by cascading water running counter-flow to the gas. This arrangement could deliver 98% methane with manufacturers guaranteeing maximum 2% methane loss in the system. It takes roughly between 3% and 6% of the total energy output in gas to run a biogas upgrading system.

Biogas Gas-grid Injection

Gas-grid injection is the injection of biogas into the methane grid (natural gas grid). Injections includes biogas until the breakthrough of micro combined heat and power two-thirds of all the energy produced by biogas power plants was lost (the heat), using the grid to transport the gas to customers, the electricity and the heat can be used for on-site generation resulting in a reduction of losses in the transportation of energy. Typical energy losses in natural gas transmission systems range from 1% to 2%. The current energy losses on a large electrical system range from 5% to 8%.

Biogas in Transport

"Biogaståget Amanda" ("The Biogas Train Amanda") train near Linköping station, Sweden.

If concentrated and compressed, it can be used in vehicle transportation. Compressed biogas is becoming widely used in Sweden, Switzerland, and Germany. A biogas-powered train, named Biogaståget Amanda (The Biogas Train Amanda), has been in service in Sweden since 2005. Biogas powers automobiles. In 1974, a British documentary film titled *Sweet as a Nut* detailed the biogas production process from pig manure and showed how it fueled a custom-adapted combustion engine. In 2007, an estimated 12,000 vehicles were being fueled with upgraded biogas worldwide, mostly in Europe.

Measuring in Biogas Environments

Biogas is part of the wet gas and condensing gas (or air) category that includes mist or fog in the gas stream. The mist or fog is predominately water vapor that condenses on the sides of pipes or stacks throughout the gas flow. Biogas environments include wastewater digesters, landfills, and animal feeding operations (covered livestock lagoons).

Ultrasonic flow meters are one of the few devices capable of measuring in a biogas atmosphere. Most thermal flow meters are unable to provide reliable data because the moisture causes steady high flow readings and continuous flow spiking, although there are single-point insertion thermal mass flow meters capable of accurately monitoring biogas flows with minimal pressure drop. They can handle moisture variations that occur in the flow stream because of daily and seasonal temperature fluctuations, and account for the moisture in the flow stream to produce a dry gas value.

Legislation

European Union

The European Union has legislation regarding waste management and landfill sites called the Landfill Directive.

Countries such as the United Kingdom and Germany now have legislation in force that provides farmers with long-term revenue and energy security.

United States

The United States legislates against landfill gas as it contains VOCs. The United States Clean Air Act and Title 40 of the Code of Federal Regulations (CFR) requires landfill owners to estimate the quantity of non-methane organic compounds (NMOCs) emitted. If the estimated NMOC emissions exceeds 50 tonnes per year, the landfill owner is required to collect the gas and treat it to remove the entrained NMOCs. Treatment of the landfill gas is usually by combustion. Because of the remoteness of landfill sites, it is sometimes not economically feasible to produce electricity from the gas.

Global Developments

United States

With the many benefits of biogas, it is starting to become a popular source of energy and is starting to be used in the United States more. In 2003, the United States consumed 147 trillion BTU of energy from "landfill gas", about 0.6% of the total U.S. natural gas consumption. Methane biogas derived from cow manure is being tested in the U.S. According to a 2008 study, collected by the *Science and Children* magazine, methane biogas from cow manure would be sufficient to produce 100 billion kilowatt hours enough to power millions of homes across America. Furthermore, methane biogas has been tested to prove that it can reduce 99 million metric tons of greenhouse gas emissions or about 4% of the greenhouse gases produced by the United States.

In Vermont, for example, biogas generated on dairy farms was included in the CVPS Cow Power program. The program was originally offered by Central Vermont Public Service Corporation as a voluntary tariff and now with a recent merger with Green Mountain Power is now the GMP Cow Power Program. Customers can elect to pay a premium on their electric bill, and that premium is passed directly to the farms in the program. In Sheldon, Vermont, Green Mountain Dairy has provided renewable energy as part of the Cow Power program. It started when the brothers who own the farm, Bill and Brian Rowell, wanted to address some of the manure management challenges faced by dairy farms, including manure odor, and nutrient availability for the crops they need to grow to feed the animals. They installed an anaerobic digester to process the cow and milking center waste from their 950 cows to produce renewable energy, a bedding to replace sawdust, and a plant-friendly fertilizer. The energy and environmental attributes are sold to the GMP Cow Power program. On average, the system run by the Rowells produces enough electricity to power 300 to 350 other homes. The generator capacity is about 300 kilowatts.

In Hereford, Texas, cow manure is being used to power an ethanol power plant. By switching to methane biogas, the ethanol power plant has saved 1000 barrels of oil a day. Over all, the power plant has reduced transportation costs and will be opening many more jobs for future power plants that will rely on biogas.

In Oakley, Kansas, an ethanol plant considered to be one of the largest biogas facilities in North America is using Integrated Manure Utilization System "IMUS" to produce heat for its boilers by utilizing feedlot manure, municipal organics and ethanol plant waste. At full capacity the plant is expected to replace 90% of the fossil fuel used in the manufacturing process of ethanol.

Europe

The level of development varies greatly in Europe. While countries such as Germany, Austria and Sweden are fairly advanced in their use of biogas, there is a vast potential for this renewable energy source in the rest of the continent, especially in Eastern Europe. Different legal frameworks, education schemes and the availability of technology are among the prime reasons behind this untapped potential. Another challenge for the further progression of biogas has been negative public perception.

Initiated by the events of the gas crisis in Europe during December 2008, it was decided to launch the EU project "SEBE" (Sustainable and Innovative European Biogas Environment) which is financed under the CENTRAL programme. The goal is to address the energy dependence of Europe by establishing an online platform to combine knowledge and launch pilot projects aimed at raising awareness among the public and developing new biogas technologies.

In February 2009, the European Biogas Association (EBA) was founded in Brussels as a non-profit organisation to promote the deployment of sustainable biogas production and use in Europe. EBA's strategy defines three priorities: establish biogas as an important part of Europe's energy mix, promote source separation of household waste to increase the gas potential, and support the production of biomethane as vehicle fuel. In July 2013, it had 60 members from 24 countries across Europe.

UK

As of September 2013, there are about 130 non-sewage biogas plants in the UK. Most are on-farm, and some larger facilities exist off-farm, which are taking food and consumer wastes.

On 5 October 2010, biogas was injected into the UK gas grid for the first time. Sewage from over 30,000 Oxfordshire homes is sent to Didcot sewage treatment works, where it is treated in an anaerobic digestor to produce biogas, which is then cleaned to provide gas for approximately 200 homes.

In 2015 the Green-Energy company Ecotricity announced their plans to build three grid-injecting digester's.

Germany

Germany is Europe's biggest biogas producer and the market leader in biogas technology. In 2010 there were 5,905 biogas plants operating throughout the country: Lower Saxony, Bavaria, and the eastern federal states are the main regions. Most of these plants are employed as power plants. Usually the biogas plants are directly connected with a CHP which produces electric power by burning the bio methane. The electrical power is then fed into the public power grid. In 2010, the total installed electrical capacity of these power plants was 2,291 MW. The electricity supply was approximately 12.8 TWh, which is 12.6% of the total generated renewable electricity.

Biogas in Germany is primarily extracted by the co-fermentation of energy crops (called 'NawaRo', an abbreviation of *nachwachsende Rohstoffe*, German for renewable resources) mixed with manure. The main crop used is corn. Organic waste and industrial and agricultural residues such as waste from the food industry are also used for biogas generation.In this respect, biogas production in Germany differs significantly from the UK, where biogas generated from landfill sites is most common.

Biogas production in Germany has developed rapidly over the last 20 years. The main reason is the legally created frameworks. Government support of renewable energy started in 1991 with the Electricity Feed-in Act (*StrEG*). This law guaranteed the producers of energy from renewable sources the feed into the public power grid, thus the power companies were forced to take all produced energy from independent private producers of green energy. In 2000 the Electricity Feed-in Act was replaced by the Renewable Energy Sources Act (*EEG*). This law even guaranteed a fixed compensation for the produced electric power over 20 years. The amount of around 8 ¢/kWh gave farmers the opportunity to become energy suppliers and gain a further source of income.

The German agricultural biogas production was given a further push in 2004 by implementing the so-called NawaRo-Bonus. This is a special payment given for the use of renewable resources, that is, energy crops. In 2007 the German government stressed its intention to invest further effort and support in improving the renewable energy supply to provide an answer on growing climate challenges and increasing oil prices by the 'Integrated Climate and Energy Programme'.

This continual trend of renewable energy promotion induces a number of challenges facing the management and organisation of renewable energy supply that has also several impacts on the biogas production. The first challenge to be noticed is the high area-consuming of the biogas electric power supply. In 2011 energy crops for biogas production consumed an area of circa 800,000 ha in Germany. This high demand of agricultural areas generates new competitions with the food industries that did not exist hitherto. Moreover, new industries and markets were created in predominately rural regions entailing different new players with an economic, political and civil background. Their influence and acting has to be governed to gain all advantages this new source of energy is offering. Finally biogas will furthermore play an important role in the German renewable energy supply if good governance is focused.

Indian Subcontinent

Biogas in India has been traditionally based on dairy manure as feed stock and these "gobar" gas plants have been in operation for a long period of time, especially in rural India. In the last 2-3 decades, research organisations with a focus on rural energy security have enhanced the design of the systems resulting in newer efficient low cost designs such as the Deenabandhu model.

The Deenabandhu Model is a new biogas-production model popular in India. (*Deenabandhu* means "friend of the helpless.") The unit usually has a capacity of 2 to 3 cubic metres. It is constructed using bricks or by a ferrocement mixture. In India, the brick model costs slightly more than the ferrocement model; however, India's Ministry of New and Renewable Energy offers some subsidy per model constructed.

LPG (Liquefied Petroleum Gas) is a key source of cooking fuel in urban India and its prices have been increasing along with the global fuel prices. Also the heavy subsidies provided by the successive governments in promoting LPG as a domestic cooking fuel has become a financial burden renewing the focus on biogas as a cooking fuel alternative in urban establishments. This has led to the development of prefabricated digester for modular deployments as compared to RCC and cement structures which take a longer duration to construct. Renewed focus on process technology like the Biourja process model has enhanced the stature of medium and large scale anaerobic digester in India as a potential alternative to LPG as primary cooking fuel.

In India, Nepal, Pakistan and Bangladesh biogas produced from the anaerobic digestion of manure in small-scale digestion facilities is called gobar gas; it is estimated that such facilities exist in over 2 million households in India, 50,000 in Bangladesh and thousands in Pakistan, particularly North Punjab, due to the thriving population of livestock. The digester is an airtight circular pit made of concrete with a pipe connection. The manure is directed to the pit, usually straight from the cattle shed. The pit is filled with a required quantity of wastewater. The gas pipe is connected to the kitchen fireplace through control valves. The combustion of this biogas has very little odour or smoke. Owing to simplicity in implementation and use of cheap raw materials in villages, it is one of the most environmentally sound energy sources for rural needs. One type of these system is the Sintex Digester. Some designs use vermiculture to further enhance the slurry produced by the biogas plant for use as compost.

To create awareness and associate the people interested in biogas, the Indian Biogas Association was formed. It aspires to be a unique blend of nationwide operators, manufacturers and planners of biogas plants, and representatives from science and research. The association was founded in 2010 and is now ready to start mushrooming. Its motto is "propagating Biogas in a sustainable way".

In Pakistan, the Rural Support Programmes Network is running the Pakistan Domestic Biogas Programme which has installed 5,360 biogas plants and has trained in excess of 200 masons on the technology and aims to develop the Biogas Sector in Pakistan.

In Nepal, the government provides subsidies to build biogas plant.

China

The Chinese had experimented the applications of biogas since 1958. Around 1970, China had installed 6,000,000 digesters in an effort to make agriculture more efficient. During the last years the technology has met high growth rates. This seems to be the earliest developments in generating biogas from agricultural waste.

In Developing Nations

Domestic biogas plants convert livestock manure and night soil into biogas and slurry, the fermented manure. This technology is feasible for small-holders with livestock producing 50 kg manure per day, an equivalent of about 6 pigs or 3 cows. This manure has to be collectable to mix it with water and feed it into the plant. Toilets can be connected. Another precondition is the temperature that affects the fermentation process. With an optimum at 36 C° the technology especially applies for those living in a (sub) tropical climate. This makes the technology for small holders in developing countries often suitable.

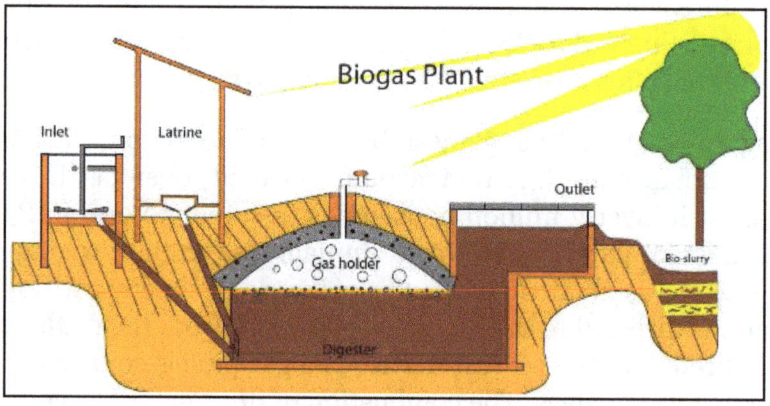

Simple sketch of household biogas plant.

Depending on size and location, a typical brick made fixed dome biogas plant can be installed at the yard of a rural household with the investment between US$300 to $500 in Asian countries and up to $1400 in the African context. A high quality biogas plant needs minimum maintenance costs and can produce gas for at least 15–20 years without major problems and re-investments. For the user, biogas provides clean cooking energy, reduces indoor air pollution, and reduces the time needed for traditional biomass collection, especially for women and children. The slurry is a clean organic fertilizer that potentially increases agricultural productivity.

Domestic biogas technology is a proven and established technology in many parts of the world, especially Asia. Several countries in this region have embarked on large-scale programmes on domestic biogas, such as China and India.

The Netherlands Development Organisation, SNV, supports national programmes on domestic biogas that aim to establish commercial-viable domestic biogas.

sectors in which local companies market, install and service biogas plants for households. In Asia, SNV is working in Nepal, Vietnam, Bangladesh, Bhutan, Cambodia, Lao PDR, Pakistan and Indonesia, and in Africa; Rwanda, Senegal, Burkina Faso, Ethiopia, Tanzania, Uganda, Kenya, Benin and Cameroon.

In South Africa a prebuilt Biogas system is manufactured and sold. One key feature is that installation requires less skill and is quicker to install as the digester tank is premade plastic.

Society and Culture

In the 1985 Australian film Mad Max Beyond Thunderdome the post-apocalyptic settlement Barter town is powered by a central biogas system based upon a piggery. As well as providing electricity, methane is used to power Barter's vehicles.

"Cow Town", written in the early 1940s, discuss the travails of a city vastly built on cow manure and the hardships brought upon by the resulting methane biogas. Carter McCormick, an engineer from a town outside the city, is sent in to figure out a way to utilize this gas to help power, rather than suffocate, the city.

Biochar

A piece of biochar.

Biochar is charcoal used as a soil amendment. Like most charcoal, biochar is made from biomass via pyrolysis. Biochar is under investigation as an approach to carbon sequestration to produce negative carbon dioxide emissions. Biochar thus has the potential to help mitigate climate change via carbon sequestration. Independently, biochar can increase soil fertility of acidic soils (low pH soils), increase agricultural productivity, and provide protection against some foliar and soil-borne diseases. Furthermore, biochar reduces pressure on forests. Biochar is a stable solid, rich in carbon, and can endure in soil for thousands of years.

History

Pre-Columbian Amazonians are believed to have used biochar to enhance soil productivity. They produced it by smoldering agricultural waste (i.e., covering burning biomass with soil) in pits or trenches. European settlers called it terra preta de Indio.

Following observations and experiments, a research team working in French Guiana hypothesized that the Amazonian earthworm *Pontoscolex corethrurus* was the main agent of fine powdering and incorporation of charcoal debris to the mineral soil.

The term "biochar" was coined by Peter Read to describe charcoal used as a soil improvement.

Production

Biochar is a high-carbon, fine-grained residue that today is produced through modern pyrolysis processes, which is the direct thermal decomposition of biomass in the absence of oxygen, which prevents combustion, to obtain an array of solid (biochar), liquid (bio-oil), and gas (syngas) products. The specific yield from the pyrolysis is dependent on process conditions. such as temperature, and can be optimized to produce either energy or biochar. Temperatures of 400–500 °C (752–932 °F) produce more char, while temperatures above 700 °C (1,292 °F) favor the yield of liquid and gas fuel components. Pyrolysis occurs more quickly at the higher temperatures, typically requiring seconds instead of hours. High temperature pyrolysis is also known as gasification, and produces primarily syngas. Typical yields are 60% bio-oil, 20% biochar, and 20% syngas. By comparison, slow pyrolysis can produce substantially more char (~50%). Once initialized, both processes produce net energy. For typical inputs, the energy required to run a "fast" pyrolyzer is approximately 15% of the energy that it outputs. Modern pyrolysis plants can use the syngas created by the pyrolysis process and output 3–9 times the amount of energy required to run.

The Amazonian pit/trench method harvests neither bio-oil nor syngas, and releases a large amount of CO_2, black carbon, and other greenhouse gases (GHG)s (and potentially, toxins) into the air. Commercial-scale systems process agricultural waste, paper byproducts, and even municipal waste and typically eliminate these side effects by capturing and using the liquid and gas products.

Centralized, Decentralized, and Mobile Systems

In a centralized system, all biomass in a region is brought to a central plant for processing. Alternatively, each farmer or group of farmers can operate a lower-tech kiln. Finally, a truck equipped with a pyrolyzer can move from place to place to pyrolyze biomass. Vehicle power comes from the syngas stream, while the biochar remains on the farm. The biofuel is sent to a refinery or storage site. Factors that influence the choice of system type include the cost of transportation of the liquid and solid byproducts, the amount of material to be processed, and the ability to feed directly into the power grid.

For crops that are not exclusively for biochar production, the residue-to-product ratio (RPR) and the collection factor (CF) the percent of the residue not used for other things, measure the approximate amount of feedstock that can be obtained for pyrolysis after harvesting the primary product. For instance, Brazil harvests approximately 460 million tons (MT) of sugarcane annually, with an RPR of 0.30, and a CF of 0.70 for the sugarcane tops, which normally are burned in the field. This translates into approximately 100 MT of residue annually, which could be pyrolyzed to create energy and soil additives. Adding in the bagasse (sugarcane waste) (RPR=0.29 CF=1.0), which is otherwise burned (inefficiently) in boilers, raises the total to 230 MT of pyrolysis feedstock. Some plant residue, however, must remain on the soil to avoid increased costs and emissions from nitrogen fertilizers.

Pyrolysis technologies for processing loose and leafy biomass produce both biochar and syngas.

Thermo-catalytic Depolymerization

Alternatively, "thermo-catalytic depolymerization", which utilizes microwaves, has recently been used to efficiently convert organic matter to biochar on an industrial scale, producing ~50% char.

Uses

Carbon Sink

The burning and natural decomposition of biomass and in particular agricultural waste adds large amounts of CO2 to the atmosphere. Biochar that is stable, fixed, and 'recalcitrant' carbon can store large amounts of greenhouse gases in the ground for centuries, potentially reducing or stalling the growth in atmospheric greenhouse gas levels; at the same time its presence in the earth can improve water quality, increase soil fertility, raise agricultural productivity, and reduce pressure on old-growth forests.

Biochar can sequester carbon in the soil for hundreds to thousands of years, like coal. Such a carbon-negative technology would lead to a net withdrawal of CO_2 from the atmosphere, while producing and consuming energy. This technique is advocated by prominent scientists such as James Hansen, head of the NASA Goddard Institute for Space Studies, and James Lovelock, creator of the Gaia hypothesis, for mitigation of global warming by greenhouse gas remediation.

Researchers have estimated that sustainable use of biocharring could reduce the global net emissions of carbon dioxide (CO2), methane, and nitrous oxide by up to 1.8 Pg CO2-C equivalent (CO2-C_e) per year (12% of current anthropogenic CO2-Ce emissions; 1 Pg=1 Gt), and total net emissions over the course of the next century by 130 Pg CO2-C_e, without endangering food security, habitat, or soil conservation.

Soil Amendment

Biochar is recognised as offering a number of benefits for soil health. Many benefits are related to the extremely porous nature of biochar. This structure is found to be very effective at retaining both water and water-soluble nutrients. Soil biologist Elaine Ingham indicates the extreme suitability of biochar as a habitat for many beneficial soil micro organisms. She points out that when pre charged with these beneficial organisms biochar becomes an extremely effective soil amendment promoting good soil, and in turn plant, health.

Biochar has also been shown to reduce leaching of E-coli through sandy soils depending on application rate, feedstock, pyrolysis temperature, soil moisture content, soil texture, and surface properties of the bacteria.

For plants that require high potash and elevated pH, biochar can be used as a soil amendment to improve yield.

Biochar can improve water quality, reduce soil emissions of greenhouse gases, reduce nutrient leaching, reduce soil acidity, and reduce irrigation and fertilizer requirements. Biochar was also found under certain circumstances to induce plant systemic responses to foliar fungal diseases and to improve plant responses to diseases caused by soilborne pathogens.

The various impacts of biochar can be dependent on the properties of the biochar, as well as the amount applied, and there is still a lack of knowledge about the important mechanisms and properties. Biochar impact may depend on regional conditions including soil type, soil condition (depleted or healthy), temperature, and humidity. Modest additions of biochar to soil reduce nitrous oxide N2O emissions by up to 80% and eliminate methane emissions, which are both more potent greenhouse gases than CO_2.

Studies have reported positive effects from biochar on crop production in degraded and nutrient–poor soils. Biochar can be designed with specific qualities to target distinct properties of soils. Biochar reduces leaching of critical nutrients, creates a higher crop uptake of nutrients, and provides greater soil availability of nutrients. At 10% levels biochar reduced contaminant levels in plants by up to 80%, while reducing total chlordane and DDX content in the plants by 68 and 79%, respectively. On the other hand, because of its high adsorption capacity, biochar may reduce the efficacy of soil applied pesticides that are needed for weed and pest control. High-surface-area biochars may be particularly problematic in this regard; more research into the long-term effects of biochar addition to soil is needed.

Slash-and-char

Switching from slash-and-burn to slash-and-char farming techniques in Brazil can decrease both deforestation of the Amazon basin and carbon dioxide emission, as well as increase crop yields. Slash-and-burn leaves only 3% of the carbon from the organic material in the soil.

Slash-and-char can keep up to 50% of the carbon in a highly stable form. Returning the biochar into the soil rather than removing it all for energy production reduces the need for nitrogen fertilizers, thereby reducing cost and emissions from fertilizer production and transport. Additionally, by improving the soil's ability to be tilled, fertility, and productivity, biochar–enhanced soils can indefinitely sustain agricultural production, whereas non-enriched soils quickly become depleted of nutrients, forcing farmers to abandon the fields, producing a continuous slash and burn cycle and the continued loss of tropical rainforest. Using pyrolysis to produce bio-energy also has the added benefit of not requiring infrastructure changes the way processing biomass for cellulosic ethanol does. Additionally, the biochar produced can be applied by the currently used machinery for tilling the soil or equipment used to apply fertilizer.

Water Retention

Biochar is a desirable soil material in many locations due to its ability to attract and retain water. This is possible because of its porous structure and high surface area. As a result, nutrients, phosphorus, and agrochemicals are retained for the plants benefit. Plants therefore, are healthier and fertilizers leach less into surface or groundwater.

Energy Production: Bio-oil and Syngas

Mobile pyrolysis units can be used to lower the costs of transportation of the biomass if the biochar is returned to the soil and the syngas stream is used to power the process. Bio-oil contains organic acids that are corrosive to steel containers, has a high water vapor content that is detrimental to ignition, and, unless carefully cleaned, contains some biochar particles which can block injectors.

If biochar is used for the production of energy rather than as a soil amendment, it can be directly substituted for any application that uses coal. Pyrolysis also may be the most cost-effective way of electricity generation from biomaterial.

Direct and Indirect Benefits

- The pyrolysis of forest- or agriculture-derived biomass residue generates a biofuel without competition with crop production.
- Biochar is a pyrolysis byproduct that may be ploughed into soils in crop fields to enhance their fertility and stability, and for medium- to long-term carbon sequestration in these soils.
- Biochar enhances the natural process: the biosphere captures CO2, especially through plant production, but only a small portion is stably sequestered for a relatively long time (soil, wood, etc.).
- Biomass production to obtain biofuels and biochar for carbon sequestration in

the soil is a carbon-negative process, i.e. more CO2 is removed from the atmosphere than released, thus enabling long-term sequestration.

Research

Intensive research into manifold aspects involving the pyrolysis/biochar platform is underway around the world. From 2005 to 2012, there were 1,038 articles that included the word "biochar" or "bio-char" in the topic that had been indexed in the ISI Web of Science. Further research is in progress by such diverse institutions around the world as Cornell University, the University of Edinburgh, which has a dedicated research unit., and the Agricultural Research Organization (ARO) of Israel, Volcani Center, where a network of researchers involved in biochar research (iBRN, Israel Biochar Researchers Network) was established as early as 2009.

Students at Stevens Institute of Technology in New Jersey are developing supercapacitors that use electrodes made of biochar. A process developed by University of Florida researchers that removes phosphate from water, also yields methane gas usable as fuel and phosphate-laden carbon suitable for enriching soil.

Emerging Commercial Sector

Calculations suggest that emissions reductions can be 12 to 84% greater if biochar is put back into the soil instead of being burned to offset fossil-fuel use. Thus biochar sequestration offers the chance to turn bioenergy into a carbon-negative industry.

Johannes Lehmann, of Cornell University, estimates that pyrolysis can be cost-effective for a combination of sequestration and energy production when the cost of a CO_2 ton reaches $37. As of mid-February 2010, CO_2 is trading at $16.82/ton on the European Climate Exchange (ECX), so using pyrolysis for bioenergy production may be feasible even if it is more expensive than fossil fuel.

Current biochar projects make no significant impact on the overall global carbon budget, although expansion of this technique has been advocated as a geoengineering approach. In May 2009, the Biochar Fund received a grant from the Congo Basin Forest Fund for a project in Central Africa to simultaneously slow down deforestation, increase the food security of rural communities, provide renewable energy and sequester carbon.

Application rates of 2.5–20 tonnes per hectare (1.0–8.1 t/acre) appear to be required to produce significant improvements in plant yields. Biochar costs in developed countries vary from $300–7000/tonne, generally too high for the farmer/horticulturalist and prohibitive for low-input field crops. In developing countries, constraints on agricultural biochar relate more to biomass availability and production time. An alternative is to use small amounts of biochar in lower cost biochar-fertilizer complexes.

Various companies in North America, Australia, and England sell biochar or biochar production units. In England Carbon Gold supply a range of biochar-based soil improvers, composts and fertilisers for arboriculture, horticulture and turfcare as well as to home growers. In Sweden the 'Stockholm Solution' is an urban tree planting system that uses 30% biochar to support healthy growth of the urban forest. The Qatar Aspire Park now uses biochar to help trees cope with the intense heat of their summers.

At the 2009 International Biochar Conference, a mobile pyrolysis unit with a specified intake of 1,000 pounds (450 kg) was introduced for agricultural applications. The unit had a length of 12 feet and height of 7 feet (3.6 m by 2.1m).

A production unit in Dunlap, Tennessee by Mantria Corporation opened in August 2009 after testing and an initial run, was later shut down as part of a Ponzi scheme investigation.

Bioliquids

Bioliquids are liquid fuels made from biomass for energy purposes other than transport (i.e. heating and electricity).

Bioliquids are usually made from virgin or used vegetable and seed oils, like palm or soya oil. These oils are burned in a power station to create heat, which can then be used to warm homes or boil water to make steam. This steam can then be used to drive a turbine to generate electricity.

Rudolf Diesel's first public exhibition of the internal combustion engine, that was to later bear his name, ran on peanut oil.

Bioliquid Production and Use

Bioliquids have been used for many years to provide heat for homes on a small scale but now big energy providers are looking at their use on a larger scale.

A controversial plant in Bristol (UK) was recently given the go ahead despite receiving several hundred complaints. The plant will be built and operated by W4B and provide enough power for 25,000 homes.

Advantages

Bioliquids have several key advantages over other sources of renewable energy:

- Bioliquids have a high energy density.
- The technology is well established, having been used for many years.

- Can be used on demand, reacting quickly to changes in demand for power.
- Can help reduce dependency on foreign oil.
- Reduces the green house gas emissions.

Disadvantages

Many of the same problems that affect biofuels also affect bioliquids and there are various social, economic, environmental and technical issues, which have been discussed in the popular media and scientific journals. These include: the effect of moderating oil prices, the "food vs fuel" debate, poverty reduction potential, carbon emissions levels, sustainable biofuel production, deforestation and soil erosion, loss of biodiversity, impact on water resources, as well as energy balance and efficiency.

Bioliquids also have several key problems compared to other sources of renewable energy:

- Price of fuel is very variable, due to competitiveness of feedstock for other uses (e.g. soap).
- Supply chain is still very new.
- Governments, such as the EU, remained undecided on bioliquids.

Blue Carbon

Blue carbon is the carbon captured by the world's oceans and coastal ecosystems. The carbon captured by living organisms in oceans is stored in the form of biomass and sediments from mangroves, salt marshes, seagrasses and potentially algae.

Overview

Historically the ocean and terrestrial forest ecosystems have been the major natural carbon (C) sinks. New research on the role of vegetated coastal ecosystems have highlighted their potential as highly efficient C sinks, and led to the scientific recognition of the term "Blue Carbon". "Blue Carbon" designates carbon that is fixed via ocean and coastal ecosystems, rather than traditional land ecosystems, like forests. Although the ocean's vegetated habitats cover less than 0.5% of the seabed, they are responsible for more than 50%, and potentially up to 70%, of all carbon storage in ocean sediments. Mangroves, salt marshes and seagrasses make up the majority of the ocean's vegetated habitats but only equal 0.05% of the plant biomass on land. Despite their small footprint, they can store a comparable amount of carbon per year and are highly efficient carbon sinks. Seagrasses, mangroves and salt marshes can capture carbon dioxide (CO_2) from the atmosphere by sequestering the C in their underlying sediments, in underground and below-ground biomass, and in dead biomass. In plant biomass such as leaves, stems, branches or roots, blue carbon can be sequestered for years to decades, and for thousands to millions of years in underlying plant sediments. Current estimates of long-term blue carbon C burial capacity are variable, and research is ongoing. Although vegetated coastal ecosystems cover less area and have less aboveground biomass than terrestrial plants they have the potential to impact longterm C sequestration, particularly in sediment sinks. One of the main concerns with Blue Carbon is the rate of loss of these important marine ecosystems is much higher than any other ecosystem on the planet, even compared to rainforests. Current estimates suggest a loss of 2-7% per year, which is not only lost carbon sequestration, but also lost habitat that is important for managing climate, coastal protection, and health.

Types of Blue Carbon Ecosystems

Seagrass

Seagrass at Rapid Bay Jetty, South Australia.

Seagrass are a group of about 60 angiosperm species that have adapted to an aquatic life, and can grow in meadows along the shores of all continents except Antarctica. Seagrass meadows form in maximum depths of up to 50m, depending on water quality and light availability, and can include up to 12 different species in one meadow. These seagrass meadows are highly productive habitats that provide many ecosystem services, including sediment stabilization, habitat and biodiversity, better water quality, and carbon and nutrient sequestration. The current documented seagrass area is 177,000 km^2, but is thought to underestimate the total area since many areas with large seagrass meadows have not been thoroughly documented. Most common estimates are 300,000 to 600,000 km^2, with up to 4,320,000 km^2 suitable seagrass habitat worldwide. Although seagrass makes up only 0.1% of the area of the ocean floor, it accounts for approximately 10-18% of the total oceanic carbon burial. Currently global seagrass meadows are estimated to store as much as 19.9 Pg (gigaton, or billion tons) of organic carbon. Carbon primarily accumulates in marine sediments, which are anoxic and thus continually preserve organic carbon from decadal-millennial time scales. High accumulation rates, low oxygen, low sediment conductivity and slower microbial decomposition rates all encourage carbon burial and carbon accumulation in these coastal sediments. Compared to terrestrial habitats that lose carbon stocks as CO_2 during decomposition or by disturbances like fires or deforestation, marine carbon sinks can retain C for much longer time periods. Carbon sequestration rates in seagrass meadows vary depending on the species, characteristics of the sediment, and depth of the habitats, but on average the carbon burial rate is approximately 138 g C m^{-2} yr^{-1}. Seagrass habitats are threatened by coastal eutrophication, increased seawater temperatures, increased sedimentation and coastal development, and sea-level rise which may decrease light availability for photosynthesis. Seagrass loss has accelerated over the past few decades, from 0.9% per year prior to 1940 to 7% per year in 1990, with about 1/3 of global loss since WWII. Scientists encourage protection and continued research of these ecosystems for organic carbon storage, valuable habitat and other ecosystem services.

Mangrove

Mangrove forest in Everglades National Park, FL.

Mangroves are woody halophytes that form intertidal forests and provide many important ecosystem services including coastal protection, nursery grounds for coastal fish and crustaceans, forest products, recreation, nutrient filtration and carbon sequestration. Currently they are found in 123 countries, with 73 identified species. They grow along coastlines in subtropical and tropical waters, depending mainly on temperature, but also vary with precipitation, tides, waves and water flow. Because they grow at the intersection between land and sea, they have semi-terrestrial and marine components, including unique adaptations including aerial roots, viviparous embryos, and highly efficient nutrient retention mechanisms. Mangroves cover approximately 150,000 km^2 worldwide, but have declined by 20% in the last 25 years, mainly due to coastal development and land conversion. Mangrove deforestation is slowing, from 1.04% loss per year in the 1980s to 0.66% loss in the early 2000s, as research and understanding of mangrove benefits have increased. Mangrove forests are responsible for approximately 10% of global carbon burial, with an estimated carbon burial rate of 174 g C m^{-2} yr^{-1}. Mangroves, like seagrasses, have potential for high levels of carbon sequestration. They account for 3% of the global carbon sequestration by tropical forests and 14% of the global coastal ocean's carbon burial. Mangroves are naturally disturbed by floods, tsunamis, coastal storms like cyclones and hurricanes, lightning, disease and pests, and changes in water quality or temperature. Although they are resilient to many of these natural disturbances, they are highly susceptible to human impacts including urban development, aquaculture, mining, and overexploitation of shellfish, crustaceans, fish and timber. Mangroves provide globally important ecosystem services and carbon sequestration and are thus an important habitat to conserve and repair when possible.

Marsh

Tidal marsh in Hilton Head, SC.

Marshes, intertidal ecosystems dominated by herbaceous vegetation, can be found globally on coastlines from the arctic to the subtropics. In the tropics, marshes are replaced by mangroves as the dominant coastal vegetation. Marshes have high productivity, with a large portion of primary production in belowground biomass. This belowground biomass can form deposits up to 8m deep. Marshes provide valuable habitat for plants, birds, and juvenile fish, protect coastal habitat from storm surge and flooding, and can reduce nutrient loading to coastal waters. Similarly to mangrove and seagrass habitats, marshes also serve as important carbon sinks. Marshes sequester C in underground biomass due to high rates of organic sedimentation and anaerobic-dominated decomposition. Salt marshes cover approximately 22,000 to 400,000 km^2 globally, with an estimated carbon burial rate of 210 g C m^{-2} yr^{-1}. Tidal marshes have been impacted by humans for centuries, including modification for grazing, haymaking, reclamation for agriculture, development and ports, evaporation ponds for salt production, modification for aquaculture, insect control, tidal power and flood protection. Marshes are also susceptible to pollution from oil, industrial chemicals, and most commonly, eutrophication. Introduced species, sea-level rise, river damming and decreased sedimentation are additional longterm changes that affect marsh habitat, and in turn, may affect carbon sequestration potential.

Algae

Both macroalgae and microalgae are being investigated as possible means of carbon sequestration. Because algae lack the complex lignin associated with terrestrial plants, the carbon in algae is released into the atmosphere more rapidly than carbon captured on land. Algae have been proposed as a short-term storage pool of carbon that can be used as a feedstock for the production of various biogenic fuels. Microalgae are often put forth as a potential feedstock for carbon-neutral biodiesel and biomethane production due to their high lipid content. Macroalgae, on the other hand, do not have high lipid content and have limited potential as biodiesel feedstock, although they can still be used as feedstock for other biofuel generation. Macroalgae have also been investigated as a feedstock for the production of biochar. The biochar produced from macroalgae is higher in agriculturally important nutrients than biochar produced from terrestrial sources. Another novel approach to carbon capture which utilizes algae is the Bicarbonate-based Integrated Carbon Capture and Algae Production Systems (BICCAPS) developed by a collaboration between Washington State University in the United States and Dalian Ocean University in China. Many cyanobacteria, microalgae, and macroalgae species can utilize carbonate as a carbon source for photosynthesis. In the BICCAPS, alkaliphilic microalgae utilize carbon captured from flue gases in the form of bicarbonate. In South Korea, macroalgae have been utilized as part of a climate change mitigation program. The country has established the Coastal CO_2 Removal Belt (CCRB) which is composed of artificial and natural ecosystems. The goal is to capture carbon using large areas of kelp forest.

Ecosystem Restoration

Restoration of mangrove forests, seagrass meadows, marshes, and kelp forests has been implemented in many countries. These restored ecosystems have the potential to act as carbon sinks. Restored seagrass meadows were found to start sequestering carbon in sediment within about four years. This was the time needed for the meadow to reach sufficient shoot density to cause sediment deposition. Similarly, mangrove plantations in China showed higher sedimentation rates than barren land and lower sedimentation rates than established mangrove forests. This pattern in sedimentation rate is thought to be a function of the plantation's young age and lower vegetation density.

Nutrient Stoichiometry of Seagrasses

The primary nutrients determining sea grass growth are carbon (C), nitrogen (N), phosphorus (P), and light for photosynthesis. Nitrogen and P can be acquired from sediment pore water or from the water column, and sea grasses can uptake N in both ammonium (NH^{4+}) and nitrate (NO^{3-}) form.

A number of studies from around the world have found that there is a wide range in the concentrations of C, N, and P in seagrasses depending on their species and environmental factors. For instance, plants collected from high-nutrient environments had lower C:N and C:P ratios than plants collected from low-nutrient environments. Sea grass stoichiometry does not follow the Redfield ratio commonly used as an indicator of nutrient availability for phytoplankton growth. In fact, a number of studies from around the world have found that the proportion of C:N:P in sea grasses can vary significantly depending on their species, nutrient availability, or other environmental factors. Depending on environmental conditions, sea grasses can be either P-limited or N-limited.

An early study of sea grass stoichiometry suggested that the "Redfield" balanced ratio between N and P for sea grasses is approximately 30:1. However, N and P concentrations are strictly not correlated, suggesting that sea grasses can adapt their nutrient uptake based on what is available in the environment. For example, sea grasses from meadows fertilized with bird excrement have shown a higher proportion of phosphate than unfertilized meadows. Alternately, sea grasses in environments with higher loading rates and organic matter diagenesis supply more P, leading to N-limitation. P availability in *T. testudinum* is the limiting nutrient. The nutrient distribution in *T. testudinum* ranges from 29.4-43.3% C, 0.88-3.96% N, and 0.048-0.243% P. This equates to a mean ratio of 24.6 C:N, 937.4 C:P, and 40.2 N:P. This information can also be used to characterize the nutrient availability of a bay or other water body (which is difficult to measure directly) by sampling the sea grasses living there.

Light availability is another factor that can affect the nutrient stoichiometry of sea grasses. Nutrient limitation can only occur when photosynthetic energy causes grasses to grow faster than the influx of new nutrients. For example, low light environments tend to have a lower C:N ratio. Alternately, high-N environments can have an indirect negative effect to sea grass growth by promoting growth of algae that reduce the total amount of available light.

Nutrient variability in sea grasses can have potential implications for wastewater management in coastal environments. High amounts of anthropogenic nitrogen discharge could cause eutrophication in previously N-limited environments, leading to hypoxic conditions in the sea grass meadow and affecting the carrying capacity of that ecosystem.

A study of annual deposition of C, N, and P from P. Oceanica sea grass meadows in northeast Spain found that the meadow sequestered 198 g C m-2 yr-1, 13.4 g N m-2 yr-1, and 2.01 g P m-2 yr-1 into the sediment. Subsequent remineralization of carbon from the sediments due to respiration returned approximately 8% of the sequestered carbon, or 15.6 g C m-2 yr -1.

Distribution and Decline of Blue Carbon Ecosystems

Seagrasses, mangroves and marshes are types of vegetated coastal habitats that cover approximately 49 million hectares worldwide. Seagrass ecosystems range from polar to tropical regions, mangroves are found in tropical and sub-tropical ecosystems and tidal marshes are found in mostly temperate regions such as on the east coast of the United States. As habitats that sequester carbon are altered and decreased, that stored amount of C is being released into the atmosphere, continuing the current accelerated rate of climate change. Impacts on these habitats globally will directly and indirectly release the previously stored carbon, which had been sequestered in sediments of these habitats. Declines of vegetated coastal habitats are seen worldwide; examples seen in mangroves are due to clearing for shrimp ponds such is the case in Indonesia, while in seagrasses there are both natural causes due to pathogens and may be exacerbated by anthropogenic effects. Quantifying rates of decrease are difficult to calculate, however measurements have been estimated by researchers indicating that if blue carbon ecosystems continue to decline, for any number of reasons, 30-40% of tidal marshes and seagrasses and approximately 100% of mangroves could be gone in the next century.

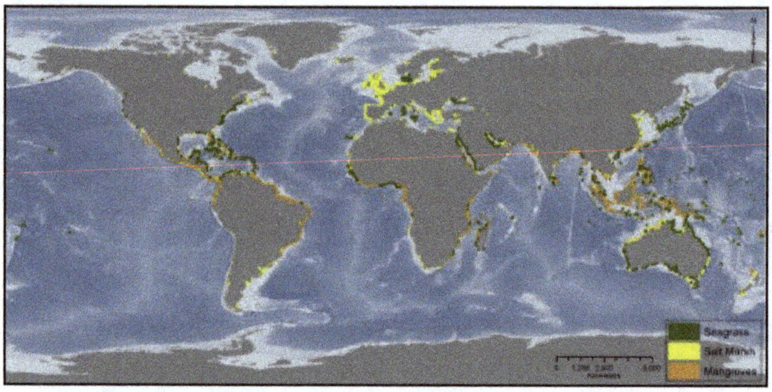

Global distribution of blue carbon.

Decline in seagrasses are due to a number of factors including drought, water quality issues, agricultural practices, invasive species, pathogens, fishing and climate change. Over 35% of global mangrove habitat remains. Decreases in habitat is due to damming of rivers, clearing for aquaculture, development etc., overfishing, and climate change, according to the World Wildlife Fund. Nearly 16% of mangroves assessed by the IUCN are on the IUCN Red List; due to development and other causes 1 in 6 worldwide mangroves are in threat of extinction. Dams threaten habitats by slowing the amount of freshwater reaching mangroves. Coral reef destruction also plays a role in mangrove habitat health as reefs slow wave energy to a level that mangroves are more tolerant of. Salt marshes may not be expansive worldwide in relation to forests, but they have a C burial rate that is over 50 times faster than tropical rainforests. Rates of burial have been estimated at up to 87.2 ± 9.6 Tg C yr-1 which is greater than that of tropical rainforests, 53 ± 9.6 Tg C yr-1. Since the 1800s salt marshes have been disturbed due to development and a lack of understanding their importance. The 25% decline since that time has led to a decrease in potential C sink area coupled with the release of once buried C. Consequences of increasingly degraded marsh habitat are a decrease in C stock in sediments, a decrease in plant biomass and thus a decrease in photosynthesis reducing the amount of CO_2 taken up by the plants, failure of C in plant blades to be transferred into the sediment, possible acceleration of erosive processes due to the lack of plant biomass, and acceleration of buried C release to the atmosphere.

Reasons for decline of mangroves, seagrass, and marshes include land use changes, climate and drought related effects, dams built in the watershed, convergence to aquaculture and agriculture, land development and sea-level rise due to climate change. Increases in these activities can lead to significant decreases in habitat available and thus increases in released C from sediments. As anthropogenic effects and climate change are heightened, the effectiveness of blue carbon sinks will diminish and CO_2 emissions will be further increased. Data on the rates at which CO_2 is being released into the atmosphere is not robust currently, however research is being conducted to gather a better information to analyze trends. Loss of underground biomass (roots and rhizomes) will allow for CO_2 to be emitted changing these habitats into sources rather than carbon sinks.

Sedimentation and Blue Carbon Burial

Organic carbon is only sequestered from the oceanic system if it reaches the sea floor and gets covered by a layer of sediment. Reduced oxygen levels in buried environments mean that tiny bacteria who eat organic matter and respire CO_2 can't decompose the carbon, so it is removed from the system permanently. Organic matter that sinks but is not buried by a sufficiently deep layer of sediment is subject to re-suspension by changing ocean currents, bioturbation by organisms that live in the top layer of marine sediments, and decomposition by heterotrophic bacteria. If any of these processes occur, the organic carbon is released back into the system. Carbon sequestration takes place only if burial rates by sediment are greater than the long term rates of erosion, bioturbation, and decomposition.

Spatial Variability in Sedimentation

Sedimentation is the rate at which floating or suspended particulate matter sinks and accumulates on the ocean floor. The faster (more energetic) the current, the more sediment it can pick up. As sediment laden currents slow, the particles fall out of suspension and come to rest on the sea floor. In other words, a fast current can pick up lots of heavy grains, where as a slow current can pick up only tiny pieces. As one can imagine, different places in the ocean vary drastically when it comes to the amount of suspended sediment and rate of deposition.

Open Ocean

The open ocean has very low sedimentation rates because most of the sediments that make it here are carried by the wind. Wind transport accounts for only a small fraction of the total sediment delivery to the oceans. Additionally, there is much less plant and animal life living in the open ocean that could be buried. Therefore, carbon burial rates are relatively slow in the open ocean.

Coastal Margins

Coastal margins have high sedimentation rates due to sediment input by rivers, which account for the vast majority of sediment delivery to the ocean. In most cases, sediments are deposited near the river mouth or are transported in the alongshore direction due to wave forcing. In some places sediment falls into submarine canyons and is transported off-shelf, if the canyon is sufficiently large or the shelf is narrow. Coastal margins also contain diverse and plentiful marine species, especially in paces that experience periodic upwelling. More marine life combined with higher sedimentation rates on coastal margins creates hotspots for carbon burial.

Submarine Canyons

Marine canyons are magnets for sediment because as currents carry sediment on the shelf in the alongshore direction, the path of the current crosses canyons perpendicularly. When the same amount of water flow is suddenly in much deeper water it slows down and deposits sediment. Due to the extreme depositional environment, carbon burial rates in the Nazare Canyon near Portugal are 30 times greater than the adjacent continental slope! This canyon alone accounts for about 0.03% of global terrestrial organic carbon burial in marine sediments. This may not seem like much, but the Nazarre submarine canyon only makes up 0.0001% of the area of the worlds ocean floor.

Human Changes to Global Sedimentary Systems

Humans have been modifying sediment cycles on a massive scale for thousands of years through a number of mechanisms.

Agriculture/Land Clearing

The first major change to global sedimentary cycling happened when humans started clearing land to grow crops. In a natural ecosystem, roots from plants hold sediment in place when it rains. Trees and shrubs reduce the amount of rainfall that impacts the dirt, and create obstacles that forest streams must flow around. When all vegetation is removed rainfall impacts directly on the dirt, there are no roots to hold on to the sediment, and there is nothing to stop the stream from scouring banks as it flows straight downhill. Because of this, land clearing causes an increase in erosion rates when compared to a natural system.

Dams

The first dams date back to 3000 BC and were built to control flood waters for agriculture. When sediment laden river flow reaches a dam's reservoir, the water slows down as it pools. Since slower water can't carry as much sediment, virtually all of the sediment falls out of suspension before the water passes through the dam. The result is that most dams are nearly 100% efficient sediment traps. Additionally, the use of dams for flood control reduces the ability of downstream channels to produce sediment. Since the vast majority of sedimentation occurs during the biggest floods, reduced frequency and intensity of flood-like flows can drastically change production rates. For thousands of years there were too few dams to have a significant impact on global sedimentary cycles, except for local impacts on a few river deltas such as the Nile which were significant. However The popularization of hydroelectric power in the last century has caused an enormous boom in dam building. Currently only a third of the world's largest rivers flow unimpeded to the ocean.

Channelization

In a natural system, the banks of a river will meander back and forth as different channels erode, accrete, open, or close. Seasonal floods regularly overwhelm riverbanks and deposit nutrients on adjacent flood plains. These services are essential to natural ecosystems, but can be troublesome for humans, who love to build infrastructure and development close to rivers. In response, rivers in populated areas are often channelized, meaning that their banks and sometimes beds are armored with a hard material, such as rocks or concrete, which prevent erosion and fixes the stream in place. This inhibits sedimentation because there is much less soft substrate left for the river to take downstream.

Impacts

Currently, the net effect of humans on global sedimentary cycling is a drastic reduction in the amount of sediment that makes it to the ocean. If we continue to build dams and channelize rivers, we will continue to see a number of problems in coastal areas including sinking deltas, shrinking beaches, and disappearing salt marshes. In addition, it's possible that we might ruin the ability of coastal margins to bury blue carbon. Without sequestration of carbon in coastal marine sediments, we will likely see accelerated global climate change.

Other Factors Influencing Blue Carbon Burial Rates

Density of Vegetation

The density of vegetation in mangrove forests, seagrass meadows, and tidal marshes is an important factor in carbon burial rates.The density of the vegetation must be sufficient to change water flows enough to reduce erosion and increase sediment deposition.

Nutrient Load

Increases in carbon capture and sequestration have been observed in both mangrove and seagrass ecosystems which have been subjected to high nutrient loads, either intentionally or due to waste from human activities. Intentional fertilization has been used in seagrass meadow restoration. Perches for seabirds are installed in the meadow and the bird droppings are the fertilizer source. The fertilization allows fast growing varieties of seagrasses to establish and grow. The species composition of these meadows is markedly different than the original seagrass meadow, although after the meadow has been reestablished and fertilization terminated, the meadows return to a species composition that more closely resembles an undisturbed meadow. Research done on mangrove soils from the Red Sea have shown that increases in nutrient loads to these soils do not increase carbon mineralization and subsequent CO_2 release. This neutral effect of fertilization was not found to be true in all mangrove forest types. Carbon capture rates also increased in these forests due to increased growth rates of the mangroves. In forests with increases in respiration there were also increases in mangrove growth of up to six times the normal rate.

Engineered Approaches to Blue Carbon

A US Department of Energy study from 2001 proposed to replicate a natural process of carbon sequestration in the ocean by combining water rich in CO_2 gas with carbonate [CO_3^-] to produce a bicarbonate [HCO_3^-] slurry. Practically, the engineered process could involve hydrating the CO_2 from power plant flue gas and running it through a porous bed of limestone to 'fix' the carbon in a saturated bicarbonate solution. This solution could then be deposited at sea to sink in the deep ocean. The cost of this process, from capture to ocean burial, was estimated to range between $90 to $180 per tonne of CO_2 and was highly dependent on the distance required to transport limestone, seawater, and the resulting bicarbonate solution.

Expected benefits from bicarbonate production over direct CO_2 gas injection would be a significantly smaller change in ocean acidity and a longer timescale for burial before the captured carbon would be released back to the atmosphere.

Cellulosic Sugars

Cellulosic sugars are derived from non-food biomass (e.g. wood, agricultural residues, municipal solid waste). The biomass is primarily composed of carbohydrate polymers cellulose, hemicellulose, and an aromatic polymer (lignin). The hemicellulose is a polymer of mainly five-carbon sugars $C_5H_{10}O_5$ (xylose). and the cellulose is a polymer of six-carbon sugar $C_6H_{12}O_6$ (glucose). Cellulose fibers are considered to be a plant's structural building blocks and are tightly bound to lignin, but the biomass can be deconstructed using Acid hydrolysis, enzymatic hydrolysis, organosolv dissolution, autohydrolysis or supercritical hydrolysis.

Biomass (cellulose, hemicellulose and lignocellulose) contain vast amounts of fermentable sugars. These sugars may be produced from a wide variety of feedstocks and can be converted into a multitude of biochemical, biofuel, and polymer products by either biological or chemical routes.

Industrial Use

In January 2012, BASF invested in Pennsylvania-based Renmatix to produce low-cost, large volume quantities of industrial sugar from lignocellulosic biomass (wood, cane bagasse or straw). Renmatix is currently the only commercial player utilizing supercritical hydrolysis as a route to cellulosic sugar production.

Renmatix is working with multiple partners on development of commercial scale facilities with the capability to produce more than 100,000 tons of cellulosic sugars annually. The company has a world-class technical center in Pennsylvania and production operations at the Integrated Plantrose Complex (IPC) in Kennesaw, Georgia and the Feedstock Processing Facility (FPF) in Rome, New York.

In June 2013, Renmatix also entered a joint development agreement (JDA) with UPM, a Finnish pulp, paper and timber manufacturer, to convert woody biomass into low-cost sugar intermediates for subsequent downstream processing into biochemicals.

In December 2013, Renmatix and Virent announced a strategic collaboration to convert affordable cellulosic sugars to renewable chemicals and bio-based packaging materials.

In March 2015, French Energy Group, Total S.A. entered a joint development agreement (JDA) with Renmatix to use the Plantrose technology to extract second-generation sugars from biomass and develop sustainable and profitable biomolecules for products of interest.

Biotechnology Penetration in the Chemical Industry

Year	Value	Penetration
2000 (actual)	$67 billion	5.3%
2005 (actual)	$98 billion	6.7%
2010 (forecast)	$159 billion	9.6%
2025 (projection)	$1000 billion	33%

World Biobased Market Penetration 2010-2025

Chemical Sector	2010	2015
Commodity Chemicals	1-2%	6-10%

Specialty Chemicals	20-25%	45-50%
Fine Chemicals	20-25%	45-50%
Polymers	5-10%	10-20%

In the first quarter of 2013, American Process Inc. announced the start-up of cellulosic sugar production using their patented AVAP® technology at their demonstration plant in Thomaston, GA. The AVAP process uses ethanol and sulfur dioxide (SO2) to fractionate biomass into it's pure components: cellulose, hemicellulose sugars and lignin. In early 2013 GranBio, a Brazilian pioneer in biofuels and biochemicals, completed the acquisition of an equity investment in API. Since that time, API has fermented both five-carbon and six-carbon sugars into high value bio-chemicals and biofuels in partnership with fermentation companies throughout the world.

Applications

Cellulosic sugars are used as renewable resources for biochemical and biofuels industries and can be used to produce intermediates by fermentative processes. The availability of industrial sugars from renewable resources, in sufficient quantities and at a favorable cost enables the products to be cost-competitive to fossil fuel based products.

A 2012 study by Nexant estimates that in the future, it will be possible and potentially economically viable to produce any type of sugar-based chemical product from biomass due to developments in cellulosic processing.

Straw

Large round bales.

Bundles of rice straw.

Straw is an agricultural by-product, the dry stalks of cereal plants, after the grain and chaff have been removed. Straw makes up about half of the yield of cereal crops such as barley, oats, rice, rye and wheat. It has many uses, including fuel, livestock bedding and fodder, thatching and basket-making. It is usually gathered and stored in a straw bale, which is a bundle of straw tightly bound with twine or wire. Bales may be square, rectangular, or round, depending on the type of baler used.

Pile of “small square” straw bales, sheltered under a tarpaulin.

Uses

Current and historic uses of straw include:

- Animal feed
    - Straw may be fed as part of the roughage component of the diet to cattle or horses that are on a near maintenance level of energy requirement. It has a low digestible energy and nutrient content. The heat generated when microorganisms in a herbivore’s gut digest straw can be useful in maintaining body temperature in cold climates. Due to the risk of impaction and its poor nutrient profile, it should always be restricted to part of the diet. It may be fed as it is, or chopped into short lengths, known as chaff.

- Basketry
    - Bee skeps and linen baskets are made from coiled and bound together continuous lengths of straw. The technique is known as lip work.
- Bedding: humans or livestock
    - The straw-filled mattress, also known as a palliasse, is still used in many parts of the world.
    - It is commonly used as bedding for ruminants and horses. It may be used as bedding and food for small animals, but this often leads to injuries to mouth, nose and eyes as straw is quite sharp.
- Biofuels
    - The use of straw as a carbon-neutral energy source is increasing rapidly, especially for biobutanol. Straw or hay briquettes are a biofuel substitute to coal.
- Biogas
    - Straw has been tested for use in a biogas plant in Aarhus University, Denmark. Straw has been processed as briquettes and feed into a biogas plant to see if higher gas yields were attainable.
- Biomass
    - The use of straw in large-scale biomass power plants is becoming mainstream in the EU, with several facilities already online. The straw is either used directly in the form of bales, or densified into pellets which allows for the feedstock to be transported over longer distances. Finally, torrefaction of straw with pelletisation is gaining attention, because it increases the energy density of the resource, making it possible to transport it still further. This processing step also makes storage much easier, because torrefied straw pellets are hydrophobic. Torrefied straw in the form of pellets can be directly co-fired with coal or natural gas at very high rates and make use of the processing infrastructures at existing coal and gas plants. Because the torrefied straw pellets have superior structural, chemical and combustion properties to coal, they can replace all coal and turn a coal plant into an entirely biomass-fed power station. First generation pellets are limited to a co-firing rate of 15% in modern IGCC plants.
- Construction material
    - In many parts of the world, straw is used to bind clay and concrete. A

mixture of clay and straw, known as cob, can be used as a building material. There are many recipes for making cob.

- When baled, straw has moderate insulation characteristics (about R-1.5/inch according to Oak Ridge National Lab and Forest Product Lab testing). It can be used, alone or in a post-and-beam construction, to build straw bale houses. When bales are used to build or insulate buildings, the straw bales are commonly finished with earthen plaster. The plastered walls provide some thermal mass, compressive and ductile structural strength, and acceptable fire resistance as well as thermal resistance (insulation), somewhat in excess of North American building code. Straw is an abundant agricultural waste product, and requires little energy to bale and transport for construction. For these reasons, strawbale construction is gaining popularity as part of passive solar and other renewable energy projects.
- Composite lumber Wheat straw can be used as a polymer filler combined with polymers to produce composite lumber.
- Enviroboard can be made from straw.
- Strawblocks.

- Crafts
  - Corn dollies.
  - Straw marquetry.
  - Straw painting.
  - Straw plaiting.
  - Scarecrows.
  - Japanese Traditional Cat's House.

Belarusian Straw Dolls.

Easter bunny made of Straw.

- Erosion control
    - Straw bales are sometimes used for sediment control at construction sites. However, bales are often ineffective in protecting water quality and are maintenance-intensive. For these reasons the U.S. Environmental Protection Agency (EPA) and various state agencies recommend use of alternative sediment control practices where possible, such as silt fences, fiber rolls and geotextiles.
    - Burned area emergency response.
    - Ground cover.
    - In-stream check dams.
- Hats
    - There are several styles of straw hats that are made of woven straw.
    - Many thousands of women and children in England (primarily in the Luton district of Bedfordshire), and large numbers in the United States (mostly Massachusetts), were employed in plaiting straw for making hats. By the late 19th century, vast quantities of plaits were being imported to England from Canton in China, and in the United States most of the straw plait was imported.
    - A fiber analogous to straw is obtained from the plant Carludovica palmata, and is used to make Panama hats.
- Horticulture
    - Straw is used in cucumber houses and for mushroom growing.
    - In Japan, certain trees are wrapped with straw to protect them from

the effects of a hard winter as well as to use them as a trap for parasite insects.

- It is also used in ponds to reduce algae by changing the nutrient ratios in the water.
- The soil under strawberries is covered with straw to protect the ripe berries from dirt, and straw is also used to cover the plants during winter to prevent the cold from killing them.
- Straw also makes an excellent mulch.

- Packaging
    - Straw is resistant to being crushed and therefore makes a good packing material. A company in France makes a straw mat sealed in thin plastic sheets.
    - Straw envelopes for wine bottles have become rarer, but are still to be found at some wine merchants.
    - Wheat straw is also used in compostable food packaging such as compostable plates. Packaging made from wheat straw can be certified compostable and will biodegrade in a commercial composting environment.
- Paper
    - Straw can be pulped to make paper.

Belarusian Straw Bird.

- Rope
    - Rope made from straw was used by thatchers, in the packaging industry and even in iron foundries.

- Shoes
    - Koreans wear Jipsin, sandals made of straw.
    - In some parts of Germany like Black Forest and Hunsrück people wear straw shoes at home or at carnival.
- Targets
    - Heavy gauge straw rope is coiled and sewn tightly together to make archery targets. This is no longer done entirely by hand, but is partially mechanised. Sometimes a paper or plastic target is set up in front of straw bales, which serve to support the target and provide a safe backdrop.
- Thatching
    - Thatching uses straw, reed or similar materials to make a waterproof, lightweight roof with good insulation properties. Straw for this purpose (often wheat straw) is grown specially and harvested using a reaper-binder.

Safety

Dried straw presents a fire hazard that can ignite easily if exposed to sparks or an open flame. It can also trigger Allergic rhinitis in people who are hypersensitive to airborne allergens such as straw dust.

Research

In addition to its current and historic uses, straw is being investigated as a source of fine chemicals including alkaloids, flavonoids, lignins, phenols, and steroids.

Woodchips

Woodchips are a medium-sized solid material made by cutting, or chipping, larger pieces of wood.

Woodchips may be used as a biomass solid fuel and are raw material for producing wood pulp. They may also be used as an organic mulch in gardening, landscaping, restoration ecology, bioreactors for denitrification and mushroom cultivation. According to the different chemical and mechanical properties of the masses, the wood logs are mostly peeled, and the bark chips and the wood chips are processed in different processes. The process of making wood chips is called woodchipping and is done with a woodchipper.

Large woodchipper (Europe Chippers model C1175).
This type of machine is used to chip large pieces of wood.

Raw Materials

Woodchips waiting to be loaded at Albany Port in Western Australia.

The raw materials of wood chips can be pulpwood, wood plantations, waste wood and residuals from construction, agriculture, landscaping, logging, and sawmills and locally grown and harvested fuel crops.

Production

A woodchipper is a machine used for reducing wood to smaller pieces. There are several types of woodchippers depending of the further processing of the woodchips. For industrial use, the woodchippers are large, stationary installations.

Pulp and Paper Industry

Wood chips used for chemical pulp must be relatively uniform in size and free of bark. The optimum size varies with the wood species. It is important to avoid damage to the wood fibres as this is important for the pulp properties. For roundwood it is most common to use disk chippers. A typical size of the disk is 2.0 - 3.5 m in diameter, 10 – 25 cm in thickness and weight is up to 30 tons. The disk is fitted with 4 to 16 knives and driven with motors of ½ - 2 MW. Drum chippers are normally used for wood residuals from saw mills or other wood industry.

Methods of Conveyance

There are four potential methods to move woodchips: pneumatic, conveyor belt, hopper with direct chute, batch system (manual conveyance).

Applications

Woodchips are used primarily as a raw material for technical wood processing. In industry, processing of bark chips is often separated after peeling the logs due to different chemical properties.

Wood Pulp

Only the heartwood and sapwood are useful for making pulp. Bark contains relatively few useful fibres and is removed and used as fuel to provide steam for use in the pulp mill. Most pulping processes require that the wood be chipped and screened to provide uniform sized chips.

Mulch

Woodchipping is also used to produce landscape and garden woodchips mulch. It is used for water conservation, weed control, reducing and preventing soil erosion, and for supporting germination of native seeds and acorns in habitat revegetation-ecological restoration projects. As the ramial chipped wood decomposes it improves the soil structure, permeability, bioactivity, and nutrient availability. Woodchips when used as a mulch are at least three inches thick.

Playground Surfacing

Wood chips can be reprocessed into an extremely effective playground surfacing material, or impact-attenuation surface. When used as playground surfacing (soft fall, cushion fall, or play chip, as it is sometimes known), woodchips can be very effective in lessening the impact of falls from playground equipment. When spread to depths of one foot (30 centimeters) playground wood chips can be effective at reducing impacts in falls up to 11 feet (3 meters). Playground woodchip is also an environmentally friendly alternative to rubber type playground surfaces.

Bioreactor

Woodchip piles at the edge of a field can inhibit nitrates from running off into water tiles. They are a simple measure for farmers to reduce nitrate pollution of the watershed without them having to change their land management practice. A 2011 study showed that most of the nitrate removal was due to heterotrophic denitrification. A 2013 experiment from Ireland showed that after 70 days of startup, a woodchip pile loaded with liquid pig manure at 5 $L/m^2/day$ removed an average of 90% of nitrate in the form of ammonium after one month. A January 2015 study from Ohio State University showed very low nitrogen gas, i.e. greenhouse gas emissions from nitrate transformation under the anaerobic conditions of the wood chip bioreactor. Scientists constructed a model for water flow and nitrate removal kinetics which can be used to design denitrification beds. It is unknown if other nutrients like phosphorus or pathogens are affected by the bioreactor as well.

Fuel

Woodchips have been traditionally used as solid fuel for space heating or in energy plants to generate electric power from renewable energy. The main source of forest chips in Europe and in most of the countries have been logging residues. It is expected that the shares of stumps and roundwood will increase in the future. As of 2013 in the EU, the estimates for biomass potential for energy, available under current conditions including sustainable use of the forest as well as providing wood to the traditional forest sectors, are: 277 million m^3, for above ground biomass and 585 million m^3 for total biomass.

The newer fuel systems for heating use either woodchips or wood pellets. The advantage of woodchips is cost, the advantage of wood pellets is the controlled fuel value. The use of woodchips in automated heating systems, is based on a robust technology.

The size of the woodchips is particularly important when burning woodchip in small plants. Unfortunately there are not many standards to decide the fractions of woodchip. One standard is the GF60 which is commonly used in smaller plants, including small industries, villas, and apartment buildings. "GF60" is known as "Fine, dry, small chips". The requirements for GF60 are that the moisture is between 10–30% and the fractions of the woodchips are distributed as follows: 0–3.5mm: <8%, 3.5–30mm: <7%, 30–60mm: 80–100%, 60–100mm: <3%, 100–120mm: <2%.

The energy content in one cubic metre is normally higher than in one cubic metre wood logs, but can vary greatly depending on moisture. The moisture is decided by the handling of the raw material. If the trees is taken down in the winter and is left to dry under the summer, with tears in the bark and covered so rain can't reach to them and then is chipped in the fall the woodchip will get an moisture of approx. 20–25%. The energy content is then approx 3.5–4.5kWh/kg (~150–250 kg/cubic metre).

Coal power plants have been converted to run on woodchips, which is fairly straightforward to do, since they both use an identical steam turbine heat engine, and the cost of woodchip fuel is comparable to coal.

Solid biomass is an attractive fuel for addressing the concerns of the energy crisis and climate change, since the fuel is affordable, widely available, close to carbon neutral and thus climate-neutral in terms of carbon dioxide (CO_2), since in the ideal case only the carbondioxide which was drawn in during the tree's growth and stored in the wood is released into the atmosphere again. It is sustainable as long as crops are allowed to regrow; In most cases, biomass is not carbon neutral as wood is not regrown and the efficiency of biomass operations produce more pollutants than the processes they replace.

Waste and Emissions

Compared to the solid waste disposal problems of coal and nuclear fuels, woodchip fuel's waste disposal problems are less grave; in a study from 2001 fly ash from wood chip combustion had 28.6 mg cadmium/kg dry matter. Compared to fly ash from burning of straw, cadmium was bound more heavily, with only small amounts of cadmium leached. It was speciated as a form of cadmium oxide, cadmium silicate ($CdSiO_3$); authors noted that adding it to agricultural or forest soils in the long-term could cause a problem with accumulation of cadmium.

Like coal, wood combustion is a known source of mercury emissions, particularly in northern climates during winter. The mercury is both gaseous as elemental mercury (especially when wood pellets are burned) or mercury oxide, and solid PM2.5 particulate matter when untreated wood is used.

When wood burning is used for space heating, indoor emissions of 1,3-butadiene, benzene, formaldehyde and acetaldehyde, which are suspected or known carcinogenic compounds, are elevated. The cancer risk from these after exposure to wood smoke is estimated to be low in developed countries.

Certain techniques for burning woodchips result in the production of biochar - effectively charcoal - which can be either utilised as charcoal, or returned to the soil, since wood ash can be used as a mineral-rich plant fertilizer. The latter method can result in an effectively carbon-negative system, as well as acting as a very effective soil conditioner, enhancing water and nutrient retention in poor soils.

Automated Handling of Solid Fuel

Unlike the smooth, uniform shape of manufactured wood pellets, woodchip sizes vary and are often mixed with twigs and sawdust. This mixture has a higher probability of jamming in small feed mechanisms. Thus, sooner or later, one or more jams is likely to occur. This reduces the reliability of the system, as well as increasing maintenance costs. Despite what some pellet stove manufacturers may say, researchers who are experienced with woodchips, say they are not compatible with the 2 inch (5 cm) auger used in pellet stoves.

Micro Combined Heat and Power

Wood is occasionally used to power engines, such as steam engines, Stirling engines, and Otto engines running on woodgas. As of 2008, these systems are rare, but as technology and the need for it develops, it is likely to be more common in the future. For the time being, wood can be increasingly used for heating applications. This will reduce the demand for heating oil, and thereby allow a greater percentage of fuel oil to be used for applications such as internal combustion engines, which are less compatible with wood based fuel and other solid biomass fuels. Heating applications generally do not require refined or processed fuels, which are almost always more expensive.

Comparison to Other Fuels

Woodchips are similar to wood pellets, in that the movement and handling is more amenable to automation than cord wood, particularly for smaller systems. Woodchips are less expensive than wood pellets and are theoretically more energy efficient than pellets, because less energy is required for manufacturing, processing, and transportation; however, this assumes that they are consumed in an appropriately designed burner, and as of 2008, these are mostly only available in large systems designed for commercial or institutional use, which have been very successful in terms of performance, cost, reliability, and efficiency.

Woodchips are less expensive than cord wood, because the harvesting is faster and more highly automated. Woodchips are of greater supply, partly because all parts of a tree can be chipped, whereas small limbs and branches can require much labor convert to cord wood. Cordwood generally needs to be "seasoned" or "dry" before it can be burned cleanly and efficiently. On the other hand, woodchip systems are typically designed to cleanly and efficiently burn "green chips" with very high moisture content of 43–47% (wet basis).

Environmental Aspects

If woodchips are harvested through sustainable forestry practices, then this is considered a source of renewable energy. On the other hand, harvesting practices, such as clearcutting large areas, are highly damaging to forest ecosystems.

Theoretically, whole-tree chip harvesting does not have as high a solar energy efficiency compared to short rotation coppice; however, it can be an energy-efficient and low-cost method of harvesting. In some cases, this practice may be controversial when whole-tree harvesting may often be associated with clear cutting and perhaps other questionable forestry practices.

Waste Processing

Woodchips, and bark chips, can be used as bulking agents in industrial composting of municipal biodegradeable waste, particularly biosolids.

Woodchip biomass does not have the waste disposal issues of coal and nuclear power, since wood ash can be used directly as a mineral-rich plant fertilizer.

Forest Fire Prevention

Woodchip harvesting can be used in concert with creating man made firebreaks, which are used as barriers to the spread of wildfire. Undergrowth coppice is ideal for chipping, and larger trees may be left in place to shade the forest floor and reduce the rate of fuel accumulation.

Market Products, Supply and Demand

Currently, domestic or residential sized systems are not available in products for sale on the general market. Homemade devices have been produced, that are small-scale, clean-burning, and efficient for woodchip fuels. Much of the research activity to date, has consisted of small budget projects that are self-funded. The majority of funding for energy research has been for liquid biofuels.

United States

"Wood chip costs usually depend on such factors as the distance from the point of delivery, the type of material (such as bark, sawmill residue or whole-tree chips), demand by other markets and how the wood fuel is transported. Chips delivered directly to the (powerplant) station by truck are less expensive than those delivered ... and shipped by railcar. The range of prices is typically between US$18 to US$30 per (wet)-ton delivered."

In 2006, prices were US$15 and US$30 per wet-ton in the northeast.

In the 20 years leading up to 2008, prices have fluctuated between US$60–70/oven-dry metric ton (odmt) in the southern states, and between US$60/odmt and US$160/odmt in the Northwest.

Europe

Large woodchipper in Germany.

In several well wooded European countries (e.g. Austria, Finland, Germany, Sweden) wood chips are becoming an alternative fuel for family homes and larger buildings due to the abundant availability of wood chips, which result in low fuel costs. The European Union is promoting wood chips for energy production in the EU Forest action plan 2007-2011. The total long term potential of wood chips in the EU it is estimated to be 913 million m3.

Japan

After a long period of negative scores, the demand of wood chip for paper manufacturing started increasing again. Starting in the last quarter of 2013, orders for printing paper and card board increased before the consumption tax increase then by weakening yen, import of papers like copy paper decreases and export of paper increases, which stimulate paper production in Japan. Softwood chip prices from the United States increased by 12% compared to October 2013 and softwood chip prices from Australia increased by 7%.

Bioconversion of Biomass to Mixed Alcohol Fuels

The bioconversion of biomass to mixed alcohol fuels can be accomplished using the MixAlco process. Through bioconversion of biomass to a mixed alcohol fuel, more energy from the biomass will end up as liquid fuels than in converting biomass to ethanol by yeast fermentation.

The process involves a biological/chemical method for converting any biodegradable material (e.g., urban wastes, such as municipal solid waste, biodegradable waste, and sewage sludge, agricultural residues such as corn stover, sugarcane bagasse, cotton gin trash, manure) into useful chemicals, such as carboxylic acids (e.g., acetic, propionic, butyric acid), ketones (e.g., acetone, methyl ethyl ketone, diethyl ketone) and biofuels, such as a mixture of primary alcohols (e.g., ethanol, propanol, *n*-butanol) and/or a mixture of secondary alcohols (e.g., isopropanol, 2-butanol, 3-pentanol). Because of the many products that can be economically produced, this process is a true biorefinery.

Pilot Plant (College Station, Texas).

The process uses a mixed culture of naturally occurring microorganisms found in natural habitats such as the rumen of cattle, termite guts, and marine and terrestrial swamps to anaerobically digest biomass into a mixture of carboxylic acids produced during the acidogenic and acetogenic stages of anaerobic digestion, however with the inhibition of the methanogenic final stage. The more popular methods for production of ethanol and cellulosic ethanol use enzymes that must be isolated first to be added to the biomass and thus convert the starch or cellulose into simple sugars, followed then by yeast fermentation into ethanol. This process does not need the addition of such enzymes as these microorganisms make their own.

As the microoganisms anaerobically digest the biomass and convert it into a mixture of carboxylic acids, the pH must be controlled. This is done by the addition of a buffering agent (e.g., ammonium bicarbonate, calcium carbonate), thus yielding a mixture of carboxylate salts. Methanogenesis, being the natural final stage of anaerobic digestion, is inhibited by the presence of the ammonium ions or by the addition of an inhibitor (e.g., iodoform). The resulting fermentation broth contains the produced carboxylate salts that must be dewatered. This is achieved efficiently by vapor-compression evaporation. Further chemical refining of the dewatered fermentation broth may then take place depending on the final chemical or biofuel product desired.

The condensed distilled water from the vapor-compression evaporation system is recycled back to the fermentation. On the other hand, if raw sewage or other waste water with high BOD in need of treatment is used as the water for the fermentation, the condensed distilled water from the evaporation can be recycled back to the city or to the original source of the high-BOD waste water. Thus, this process can also serve as a water treatment facility, while producing valuable chemicals or biofuels.

Because the system uses a mixed culture of microorganisms, besides not needing any enzyme addition, the fermentation requires no sterility or aseptic conditions, making this front step in the process more economical than in more popular methods for the production of cellulosic ethanol. These savings in the front end of the process, where volumes are large, allows flexibility for further chemical transformations after dewatering, where volumes are small.

Carboxylic Acids

Carboxylic acids can be regenerated from the carboxylate salts using a process known as "acid springing". This process makes use of a high-molecular-weight tertiary amine (e.g., trioctylamine), which is switched with the cation (e.g., ammonium or calcium). The resulting amine carboxylate can then be thermally decomposed into the amine itself, which is recycled, and the corresponding carboxylic acid. In this way, theoretically, no chemicals are consumed or wastes produced during this step.

Ketones

There are two methods for making ketones. The first one consists on thermally converting calcium carboxylate salts into the corresponding ketones. This was a common method for making acetone from calcium acetate during World War I. The other method for making ketones consists on converting the vaporized carboxylic acids on a catalytic bed of zirconium oxide.

Alcohols

Primary Alcohols

The undigested residue from the fermentation may be used in gasification to make hydrogen (H_2). This H_2 can then be used to hydrogenolyze the esters over a catalyst (e.g., copper chromite), which are produced by esterifying either the ammonium carboxylate salts (e.g., ammonium acetate, propionate, butyrate) or the carboxylic acids (e.g., acetic, propionic, butyric acid) with a high-molecular-weight alcohol (e.g., hexanol, heptanol). From the hydrogenolysis, the final products are the high-molecular-weight alcohol, which is recycled back to the esterification, and the corresponding primary alcohols (e.g., ethanol, propanol, butanol).

Secondary Alcohols

The secondary alcohols (e.g., isopropanol, 2-butanol, 3-pentanol) are obtained by hydrogenating over a catalyst (e.g., Raney nickel) the corresponding ketones (e.g., acetone, methyl ethyl ketone, diethyl ketone).

Drop-in Biofuels

The primary or secondary alcohols obtained as described above may undergo conversion to drop-in biofuels, fuels which are compatible with current fossil fuel infrastructure such as biogasoline, green diesel and bio-jet fuel. Such is done by subjecting the alcohols to dehydration followed by oligomerization using zeolite catalysts in a manner similar to the methanex process, which used to produce gasoline from methanol in New Zealand.

Acetic Acid Versus Ethanol

Cellulosic-ethanol manufacturing plants are bound to be net exporters of electricity because a large portion of the lignocellulosic biomass, namely lignin, remains undigested and it must be burned, thus producing electricity for the plant and excess electricity for the grid. As the market grows and this technology becomes more widespread, coupling the liquid fuel and the electricity markets will become more and more difficult.

Acetic acid, unlike ethanol, is biologically produced from simple sugars without the production of carbon dioxide:

$$C_6H_{12}O_6 \rightarrow 2\,CH_3CH_2OH + 2\,CO_2$$

(Biological production of ethanol)

$$C_6H_{12}O_6 \rightarrow 3\,CH_3COOH$$

(Biological production of acetic acid)

Because of this, on a mass basis, the yields will be higher than in ethanol fermentation. If then, the undigested residue (mostly lignin) is used to produce hydrogen by gasification, it is ensured that more energy from the biomass will end up as liquid fuels rather than excess heat/electricity.

$$3\,CH_3COOH + 6\,H_2 \rightarrow 3\,CH_3CH_2OH + 3\,H_2O$$

(Hydrogenation of acetic acid)

$$C_6H_{12}O_6 \text{ (from cellulose)} + 6\,H_2 \text{ (from lignin)} \rightarrow 3\,CH_3CH_2OH + 3\,H_2O$$

(Overall reaction)

A more comprehensive description of the economics of each of the fuels is given on the pages alcohol fuel and ethanol fuel, more information about the economics of various systems can be found on the central page biofuel.

Stage of Development

The system has been in development since 1991, moving from the laboratory scale (10 g/day) to the pilot scale (200 lb/day) in 2001. A small demonstration-scale plant (5 ton/day) has been constructed and is under operation and a 220 ton/day demonstration plant is expected in 2012.

Cogeneration

Cogeneration or combined heat and power (CHP) is the use of a heat engine or power station to generate electricity and useful heat at the same time. Trigeneration or combined cooling, heat and power (CCHP) refers to the simultaneous generation of electricity and useful heating and cooling from the combustion of a fuel or a solar heat collector.

Cogeneration is a thermodynamically efficient use of fuel. In separate production of electricity, some energy must be discarded as waste heat, but in cogeneration some of this thermal energy is put to use. All thermal power plants emit heat during electricity generation, which can be released into the natural environment through cooling towers, flue gas, or by other means. In contrast, CHP captures some or all of the by-product for heating, either very close to the plant, or—especially in Scandinavia and Eastern Europe—as hot water for district heating with temperatures ranging from approximately 80 to 130 °C. This is also called combined heat and power district heating (CHPDH). Small CHP plants are an example of decentralized energy. By-product heat at moderate temperatures (100–180 °C, 212–356 °F) can also be used in absorption refrigerators for cooling.

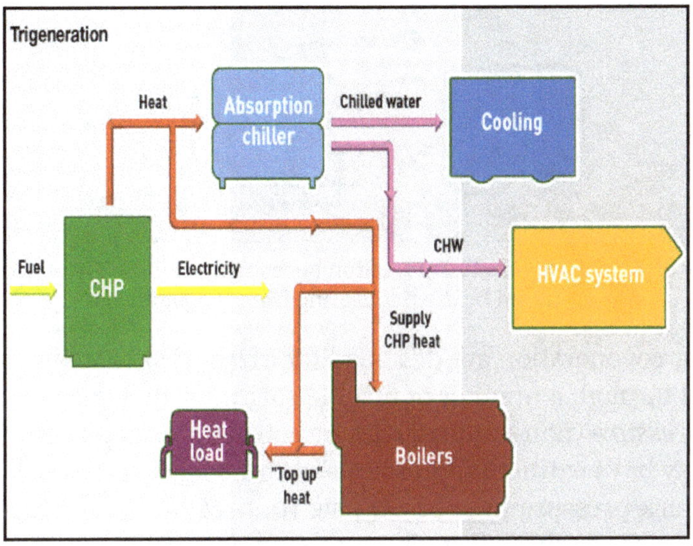

Trigeneration cycle.

The supply of high-temperature heat first drives a gas or steam turbine-powered generator and the resulting low-temperature waste heat is then used for water or space heating as described in cogeneration. At smaller scales (typically below 1 MW) a gas engine or diesel engine may be used. Trigeneration differs from cogeneration in that the waste heat is used for both heating and cooling, typically in an absorption refrigerator. CCHP systems can attain higher overall efficiencies than cogeneration or traditional power plants. In the United States, the application of trigeneration in buildings is called building cooling, heating and power (BCHP). Heating and cooling output may operate concurrently or alternately depending on need and system construction.

Cogeneration was practiced in some of the earliest installations of electrical generation. Before central stations distributed power, industries generating their own power used exhaust steam for process heating. Large office and apartment buildings, hotels and stores commonly generated their own power and used waste steam for building heat. Due to the high cost of early purchased power, these CHP operations continued for many years after utility electricity became available.

Overview

Thermal power plants (including those that use fissile elements or burn coal, petroleum, or natural gas), and heat engines in general, do not convert all of their thermal energy into electricity. In most heat engines, slightly more than half is lost as excess heat. By capturing the ex-cess heat, CHP uses heat that would be wasted in a conventional power plant, potential-ly reaching an efficiency of up to 80%, for the best conventional plants. This means that less fuel needs to be consumed to produce the same amount of useful energy.

Masnedø CHP power station in Denmark. This station burns straw as fuel. The adjacent greenhouses are heated by district heating from the plant.

Steam turbines for cogeneration are designed for *extraction* of steam at lower pressures after it has passed through a number of turbine stages, or they may be designed for final exhaust at *back pressure* (non-condensing), or both. A typical power generation turbine in a paper mill may have extraction pressures of 160 psig (1.103 MPa) and 60 psig (0.41 MPa). A typical back pressure may be 60 psig (0.41 MPa). In practice these pressures are custom designed for each facility. The extracted or exhaust steam is used for process heating, such as drying paper, evaporation, heat for chemical reactions or distillation. Steam at ordinary process heating conditions still has a considerable amount of enthalpy that could be used for power generation, so cogeneration has lost opportunity cost. Conversely, simply generating steam at process pressure instead of high enough pressure to generate power at the top end also has lost opportunity cost. The capital and operating cost of high pressure boilers, turbines and generators are substantial, and this equipment is normally operated continuously, which usually limits self-generated power to large-scale operations.

A cogeneration plant in Metz, France. The 45MW boiler uses waste wood biomass as energy source, and provides electricity and heat for 30,000 dwellings.

Some tri-cycle plants have used a combined cycle in which several thermodynamic cycles produced electricity, then a heating system was used as a condenser of the power plant's bottoming cycle. For example, the RU-25 MHD generator in Moscow heated a boiler for a conventional steam powerplant, whose condensate was then used for space heat. A more modern system might use a gas turbine powered by natural gas, whose exhaust powers a steam plant, whose condensate provides heat. Tri-cycle plants can have thermal efficiencies above 80%.

The viability of CHP (sometimes termed utilisation factor), especially in smaller CHP installations, depends on a good baseload of operation, both in terms of an on-site (or near site) electrical demand and heat demand. In practice, an exact match between the heat and electricity needs rarely exists. A CHP plant can either meet the need for heat (*heat driven operation*) or be run as a power plant with some use of its waste heat, the latter being less advantageous in terms of its utilisation factor and thus its overall efficiency. The viability can be greatly increased where opportunities for Trigeneration exist. In such cases, the heat from the CHP plant is also used as a primary energy source to deliver cooling by means of an absorption chiller.

CHP is most efficient when heat can be used on-site or very close to it. Overall efficiency is reduced when the heat must be transported over longer distances. This requires heavily insulated pipes, which are expensive and inefficient; whereas electricity can be transmitted along a comparatively simple wire, and over much longer distances for the same energy loss.

A car engine becomes a CHP plant in winter when the reject heat is useful for warming the interior of the vehicle. The example illustrates the point that deployment of CHP depends on heat uses in the vicinity of the heat engine.

Thermally enhanced oil recovery (TEOR) plants often produce a substantial amount of excess electricity. After generating electricity, these plants pump leftover steam into heavy oil wells so that the oil will flow more easily, increasing production. TEOR cogeneration plants in Kern County, California produce so much electricity that it cannot all be used locally and is transmitted to Los Angeles.

CHP is one of the most cost-efficient methods of reducing carbon emissions from heating systems in cold climates and is recognized to be the most energy efficient method of transforming energy from fossil fuels or biomass into electric power. Cogeneration plants are commonly found in district heating systems of cities, central heating systems from buildings, hospitals, prisons and are commonly used in the industry in thermal production processes for process water, cooling, steam production or CO_2 fertilization.

Types of Plants

Topping cycle plants primarily produce electricity from a steam turbine. The exhausted steam is then condensed and the low temperature heat released from this condensation is utilized for e.g. district heating or water desalination.

Bottoming cycle plants produce high temperature heat for industrial processes, then a waste heat recovery boiler feeds an electrical plant. Bottoming cycle plants are only used when the industrial process requires very high temperatures such as furnaces for glass and metal manufacturing, so they are less common.

Large cogeneration systems provide heating water and power for an industrial site or an entire town. Common CHP plant types are:

- Gas turbine CHP plants using the waste heat in the flue gas of gas turbines. The fuel used is typically natural gas.
- Gas engine CHP plants use a reciprocating gas engine which is generally more competitive than a gas turbine up to about 5 MW. The gaseous fuel used is normally natural gas. These plants are generally manufactured as fully packaged units that can be installed within a plantroom or external plant compound with simple connections to the site's gas supply, electrical distribution network and heating systems.
- Biofuel engine CHP plants use an adapted reciprocating gas engine or diesel engine, depending upon which biofuel is being used, and are otherwise very similar in design to a Gas engine CHP plant. The advantage of using a biofuel is one of reduced hydrocarbon fuel consumption and thus reduced carbon emissions. These plants are generally manufactured as fully packaged units that can be installed within a plantroom or external plant compound with simple connections to the site's electrical distribution and heating systems. Another variant is the wood gasifier CHP plant whereby a wood pellet or wood chip biofuel is gasified in a zero oxygen high temperature environment; the resulting gas is then used to power the gas engine.
- Combined cycle power plants adapted for CHP.
- Molten-carbonate fuel cells and solid oxide fuel cells have a hot exhaust, very suitable for heating.
- Steam turbine CHP plants that use the heating system as the steam condenser for the steam turbine.
- Nuclear power plants, similar to other steam turbine power plants, can be fitted with extractions in the turbines to bleed partially expanded steam to a heating system. With a heating system temperature of 95 °C it is possible to extract about 10 MW heat for every MW electricity lost. With a temperature of 130 °C the gain is slightly smaller, about 7 MW for every MWe lost.

Smaller cogeneration units may use a reciprocating engine or Stirling engine. The heat is removed from the exhaust and radiator. The systems are popular in small sizes because small gas and diesel engines are less expensive than small gas- or oil-fired steam-electric plants.

Some cogeneration plants are fired by biomass, or industrial and municipal solid waste. Some CHP plants utilize waste gas as the fuel for electricity and heat generation. Waste gases can be gas from animal waste, landfill gas, gas from coal mines, sewage gas, and combustible industrial waste gas.

Some cogeneration plants combine gas and solar photovoltaic generation to further improve technical and environmental performance. Such hybrid systems can be scaled down to the building level and even individual homes. More recent results show that solar photovoltaic + battery + CHP hybrid systems are technically viable in the continental U.S. to reduce consumer costs, while reducing energy- and electricity-related greenhouse gas emissions.

MicroCHP

Micro combined heat and power or 'Micro cogeneration" is a so-called distributed energy resource (DER). The installation is usually less than 5 kW_e in a house or small business. Instead of burning fuel to merely heat space or water, some of the energy is converted to electricity in addition to heat. This electricity can be used within the home or business or, if permitted by the grid management, sold back into the electric power grid.

Delta-ee consultants stated in 2013 that with 64% of global sales the fuel cell micro-combined heat and power passed the conventional systems in sales in 2012. 20.000 units were sold in Japan in 2012 overall within the Ene Farm project. With a Lifetime of around 60,000 hours. For PEM fuel cell units, which shut down at night, this equates to an estimated lifetime of between ten and fifteen years. For a price of $22,600 before installation. For 2013 a state subsidy for 50,000 units is in place.

The development of small-scale CHP systems has provided the opportunity for in-house power backup of residential-scale photovoltaic (PV) arrays. The results of a 2011 study show that a PV+CHP hybrid system not only has the potential to radically reduce energy waste in the status quo electrical and heating systems, but it also enables the share of solar PV to be expanded by about a factor of five. In some regions, in order to reduce waste from excess heat, an absorption chiller has been proposed to utilize the CHP-produced thermal energy for cooling of PV-CHP system. These trigeneration+-photovoltaic systems have the potential to save even more energy and further reduce emissions compared to conventional sources of power, heating and cooling.

MicroCHP installations use five different technologies: microturbines, internal combustion engines, stirling engines, closed cycle steam engines and fuel cells. One author indicated in 2008 that MicroCHP based on Stirling engines is the most cost effective of the so-called microgeneration technologies in abating carbon emissions; A 2013 UK report from Ecuity Consulting stated that MCHP is the most cost-effective method of utilising gas to generate energy at the domestic level. however, advances in reciprocation engine technology are adding efficiency to CHP plant, particularly in the biogas field. As both MiniCHP and CHP have been shown to reduce emissions they could play a large role in the field of CO_2 reduction from buildings, where more than 14% of emissions can be saved using CHP in buildings. The ability to reduce emissions is particularly strong for new communities in emission intensive grids that utilize a combination of CHP and photovoltaic systems.

Trigeneration

A plant producing electricity, heat and cold is called a trigeneration or polygeneration plant. Cogeneration systems linked to absorption chillers use waste heat for refrigeration.

Combined Heat and Power District Heating

In the United States, Consolidated Edison distributes 66 billion kilograms of 350 °F (180 °C) steam each year through its seven cogeneration plants to 100,000 buildings in Manhattan—the biggest steam district in the United States. The peak delivery is 10 million pounds per hour (or approximately 2.5 GW).

Industrial CHP

Cogeneration is still common in pulp and paper mills, refineries and chemical plants. In this "industrial cogeneration/CHP", the heat is typically recovered at higher temperatures (above 100 deg C) and used for process steam or drying duties. This is more valuable and flexible than low-grade waste heat, but there is a slight loss of power generation. The increased focus on sustainability has made industrial CHP more attractive, as it substantially reduces carbon footprint compared to generating steam or burning fuel on-site and importing electric power from the grid.

Utility Pressures Versus Self Generating Industrial

Industrial cogeneration plants normally operate at much lower boiler pressures than utilities. Among the reasons are: 1) Cogeneration plants face possible contamination of returned condensate. Because boiler feed water from cogeneration plants has much lower return rates than 100% condensing power plants, industries usually have to treat proportionately more boiler make up water. Boiler feed water must be completely oxygen free and de-mineralized, and the higher the pressure the more critical the level of purity of the feed water. 2) Utilities are typically larger scale power than industry, which helps offset the higher capital costs of high pressure. 3) Utilities are less likely to have sharp load swings than industrial operations, which deal with shutting down or starting up units that may represent a significant percent of either steam or power demand.

Heat Recovery Steam Generators

A heat recovery steam generator (HRSG) is a steam boiler that uses hot exhaust gases from the gas turbines or reciprocating engines in a CHP plant to heat up water and generate steam. The steam, in turn, drives a steam turbine or is used in industrial processes that require heat.

HRSGs used in the CHP industry are distinguished from conventional steam generators by the following main features:

- The HRSG is designed based upon the specific features of the gas turbine or reciprocating engine that it will be coupled to.
- Since the exhaust gas temperature is relatively low, heat transmission is accomplished mainly through convection.
- The exhaust gas velocity is limited by the need to keep head losses down. Thus, the transmission coefficient is low, which calls for a large heating surface area.
- Since the temperature difference between the hot gases and the fluid to be heated (steam or water) is low, and with the heat transmission coefficient being low as well, the evaporator and economizer are designed with plate fin heat exchangers.

Comparison with a Heat Pump

A heat pump may be compared with a CHP unit, in that for a condensing steam plant, as it switches to produce heat, then electrical generation becomes unavailable, just as the power used in a heat pump becomes unavailable. Typically for every unit of electrical power lost, then about 6 units of heat are made available at about 90 °C. Thus CHP has an effective Coefficient of Performance (COP) compared to a heat pump of 6. It is noteworthy that the unit for the CHP is lost at the high voltage network and therefore incurs no losses, whereas the heat pump unit is lost at the low voltage part of the network and incurs on average a 6% loss. Because the losses are proportional to the square of the current, during peak periods losses are much higher than this and it is likely that widespread (i.e. city-wide application of heat pumps) would cause overloading of the distribution and transmission grids unless they are substantially reinforced.

It is also possible to run a heat driven operation combined with a heat pump, where the excess electricity (as heat demand is the defining factor on utilization) is used to drive a heat pump. As heat demand increases, more electricity is generated to drive the heat pump, with the waste heat also heating the heating fluid.

Distributed Generation

Trigeneration has its greatest benefits when scaled to fit buildings or complexes of buildings where electricity, heating and cooling are perpetually needed. Such installations include but are not limited to: data centers, manufacturing facilities, universities, hospitals, military complexes, and schools. Localized trigeneration has addition benefits as described by distributed generation. Redundancy of power in mission critical applications, lower power usage costs and the ability to sell electrical power back to the local utility are a few of the major benefits. Even for small buildings such as individual family homes trigeneration systems provide benefits over cogeneration because of increased energy utilization. This increased efficiency can also provide significant reduced greenhouse gas emissions, particularly for new communities.

Most industrial countries generate the majority of their electrical power needs in large centralized facilities with capacity for large electrical power output. These plants have excellent economies of scale, but usually transmit electricity long distances resulting in sizable losses, negatively affect the environment. Large power plants can use cogeneration or trigeneration systems only when sufficient need exists in immediate geographic vicinity for an industrial complex, additional power plant or a city. An example of cogeneration with trigeneration applications in a major city is the New York City steam system.

Thermal Efficiency

Every heat engine is subject to the theoretical efficiency limits of the Carnot cycle. When the fuel is natural gas, a gas turbine following the Brayton cycle is typically used. Mechanical energy from the turbine drives an electric generator. The low-grade (i.e. low temperature) waste heat rejected by the turbine is then applied to space heating or cooling or to industrial processes. Cooling is achieved by passing the waste heat to an absorption chiller.

Thermal efficiency in a trigeneration system is defined as:

$$\eta_{th} \equiv \frac{W_{out}}{Q_{in}}$$

Where:

η_{th} = Thermal efficiency

W_{out} = Total work output by all systems

Q_{in} = Total heat input into the system

Typical trigeneration models have losses as in any system. The energy distribution below is represented as a percent of total input energy:

Electricity = 45%

Heat + Cooling = 40%

Heat Losses = 13%

Electrical Line Losses = 2%

Conventional central coal- or nuclear-powered power stations convert only about 33% of their input heat to electricity. The remaining 67% emerges from the turbines as low-grade waste heat with no significant local uses so it is usually rejected to the environment. These low conversion efficiencies strongly suggest that productive uses could be found for this waste heat, and in some countries these plants do collect byproduct heat that can be sold to customers.

But if no practical uses can be found for the waste heat from a central power station, e.g., due to distance from potential customers, then moving generation to where the waste heat can find uses may be of great benefit. Even though the efficiency of a small distributed electrical generator may be lower than a large central power plant, the use of its waste heat for local heating and cooling can result in an overall use of the primary fuel supply as great as 80%. This provides substantial financial and environmental benefits.

Costs

Typically, for a gas-fired plant the fully installed cost per kW electrical is around £400/kW ($577 USD), which is comparable with large central power stations.

History

Cogeneration in Europe

A cogeneration thermal power plant in Ferrera Erbognone (PV), Italy.

The EU has actively incorporated cogeneration into its energy policy via the CHP Directive. In September 2008 at a hearing of the European Parliament's Urban Lodgment Intergroup, Energy Commissioner Andris Piebalgs is quoted as saying, "security of supply really starts with energy efficiency." Energy efficiency and cogeneration are recognized in the opening paragraphs of the European Union's Cogeneration Directive 2004/08/EC. This directive intends to support cogeneration and establish a method for calculating cogeneration abilities per country. The development of cogeneration has been very uneven over the years and has been dominated throughout the last decades by national circumstances.

The European Union generates 11% of its electricity using cogeneration. However, there is large difference between Member States with variations of the energy savings between 2% and 60%. Europe has the three countries with the world's most intensive cogeneration economies: Denmark, the Netherlands and Finland. Of the 28.46 TWh of electrical power generated by conventional thermal power plants in Finland in 2012, 81.80% was cogeneration.

Other European countries are also making great efforts to increase efficiency. Germany reported that at present, over 50% of the country's total electricity demand could be provided through cogeneration. So far, Germany has set the target to double its electricity cogeneration from 12.5% of the country's electricity to 25% of the country's electricity by 2020 and has passed supporting legislation accordingly. The UK is also actively supporting combined heat and power. In light of UK's goal to achieve a 60% reduction in carbon dioxide emissions by 2050, the government has set the target to source at least 15% of its government electricity use from CHP by 2010. Other UK measures to encourage CHP growth are financial incentives, grant support, a greater regulatory framework, and government leadership and partnership.

According to the IEA 2008 modeling of cogeneration expansion for the G8 countries, the expansion of cogeneration in France, Germany, Italy and the UK alone would effectively double the existing primary fuel savings by 2030. This would increase Europe's savings from today's 155.69 Twh to 465 Twh in 2030. It would also result in a 16% to 29% increase in each country's total cogenerated electricity by 2030.

Governments are being assisted in their CHP endeavors by organizations like COGEN Europe who serve as an information hub for the most recent updates within Europe's energy policy. COGEN is Europe's umbrella organization representing the interests of the cogeneration industry.

The European public–private partnership Fuel Cells and Hydrogen Joint Undertaking Seventh Framework Programme project ene.field deploys in 2017 up 1,000 residential fuel cell Combined Heat and Power (micro-CHP) installations in 12 states. Per 2012 the first 2 installations have taken place.

Cogeneration in the United Kingdom

In the United Kingdom, the Combined Heat and Power Quality Assurance (CHPQA) scheme regulates the combined production of heat and power. CHPQA was introduced in 1996. It defines, through calculation of inputs and outputs, "Good Quality CHP" in terms of the achievement of primary energy savings against conventional separate generation of heat and electricity. Compliance with CHPQA is required for cogeneration installations to be eligible for government subsidies and tax incentives.

Cogeneration in the United States

Perhaps the first modern use of energy recycling was done by Thomas Edison. His 1882 Pearl Street Station, the world's first commercial power plant, was a combined heat and power plant, producing both electricity and thermal energy while using waste heat to warm neighboring buildings. Recycling allowed Edison's plant to achieve approximately 50 percent efficiency.

The 250 MW Kendall Cogeneration Station plant in Cambridge, Massachusetts.

By the early 1900s, regulations emerged to promote rural electrification through the construction of centralized plants managed by regional utilities. These regulations not only promoted electrification throughout the countryside, but they also discouraged decentralized power generation, such as cogeneration.

By 1978, Congress recognized that efficiency at central power plants had stagnated and sought to encourage improved efficiency with the Public Utility Regulatory Policies Act (PURPA), which encouraged utilities to buy power from other energy producers.

Diffusion

Cogeneration plants proliferated, soon producing about 8% of all energy in the United States. However, the bill left implementation and enforcement up to individual states, resulting in little or nothing being done in many parts of the country.

The United States Department of Energy has an aggressive goal of having CHP constitute 20% of generation capacity by the year 2030. Eight Clean Energy Application Centers have been established across the nation whose mission is to develop the required technology application knowledge and educational infrastructure necessary to lead "clean energy" (combined heat and power, waste heat recovery and district energy) technologies as viable energy options and reduce any perceived risks associated with their implementation. The focus of the Application Centers is to provide an outreach and technology deployment program for end users, policy makers, utilities, and industry stakeholders.

High electric rates in New England and the Middle Atlantic make these areas of the United States the most beneficial for cogeneration.

Outside of the United States, energy recycling is more common. Denmark is probably the most active energy recycler, obtaining about 55% of its energy from cogeneration and waste heat recovery. Other large countries, including Germany, Russia, and India, also obtain a much higher share of their energy from decentralized sources.

Biorefinery

A biorefinery is a facility that integrates biomass conversion processes and equipment to produce fuels, power, heat, and value-added chemicals from biomass. The biorefinery concept is analogous to today's petroleum refinery, which produce multiple fuels and products from petroleum.

The International Energy Agency Bioenergy Task 42 on Biorefineries has defined biorefining as the sustainable processing of biomass into a spectrum of bio-based products (food, feed, chemicals, materials) and bioenergy (biofuels, power and/or heat).

By producing multiple products, a biorefinery takes advantage of the various components in biomass and their intermediates therefore maximizing the value derived from the biomass feedstock. A biorefinery could, for example, produce one or several low-volume, but high-value, chemical or nutraceutical products and a low-value, but high-volume liquid transportation fuel such as biodiesel or bioethanol. At the same time generating electricity and process heat, through combined heat and power (CHP) technology, for its own use and perhaps enough for sale of electricity to the local utility. The high-value products increase profitability, the high-volume fuel helps meet energy needs, and the power production helps to lower energy costs and reduce greenhouse gas emissions from traditional power plant facilities. Although some facilities exist that can be called bio-refineries, the bio-refinery has yet to be fully realized. Future biorefineries may play a major role in producing chemicals and materials that are traditionally produced from petroleum.

Examples

The fully operational Blue Marble Energy company has multiple biorefineries located in Odessa, WA and Missoula, MT.

Canada's first Integrated Biorefinery, developed on anaerobic digestion technology by Himark BioGas is located in Hairy Hill, Alberta. The Biorefinery utilizes Source Separated Organics from the metro Edmonton region, Open Pen Feedlot Manure, and Food Processing Waste.

Several potential biorefinery examples have been proposed, starting from feedstocks such as tobacco, flax straw and the residues from the production of bioethanol.

Chemrec's technology for black liquor gasification and production of second-generation biofuels such as biomethanol or BioDME is integrated with a host pulp mill and utilizes a major sulfate or sulfite process waste product as feedstock.

Bioasphalt

Bioasphalt is an asphalt alternative made from non-petroleum based renewable resources.

These sources include sugar, molasses and rice, corn and potato starches, natural tree and gum resins, natural latex rubber and vegetable oils, lignin, cellulose, palm oil waste, coconut waste, peanut oil waste, canola oil waste, dried sewerage effluent and so on. Bitumen can also be made from waste vacuum tower bottoms produced in the process of cleaning used motor oils, which are normally burned or dumped into land fills.

Non-petroleum based bitumen binders can be colored, which can reduce the temperatures of road surfaces and reduce the Urban heat islands.

Petroleum, Environmental, and Heat Concerns

Because of concerns over Peak oil, pollution and climate change, as well the oil price increases since 2003, non-petroleum alternatives have become more popular. This has led to the introduction of biobitumen alternatives that are more environmentally friendly and nontoxic.

For millions of people living in and around cities, heat islands are of growing concern. This phenomenon describes urban and suburban temperatures that are 1 to 6 °C (2 to 10 °F) hotter than nearby rural areas. Elevated temperatures can impact communities by increasing peak energy demand, air conditioning costs, air pollution levels, and heat-related illness and mortality. Fortunately, there are common-sense measures that communities can take to reduce the negative effects of heat islands, such as replacing conventional black asphalt road surfaces with the new pigmentable bitumen that gives lighter colors.

History and Implementation

Asphalt made with vegetable oil based binders was patented by Colas SA in France in 2004.

A number of homeowners seeking an environmentally friendly alternative to asphalt for paving have experimented with waste vegetable oil as a binder for driveways and parking areas in single-family applications. The earliest known test occurred in 2002 in

Ohio, where the homeowner combined waste vegetable oil with dry aggregate to create a low-cost and less polluting paving material for his 200-foot driveway. After five years, he reports the driveway is performing as well or better than petroleum-based materials.

Shell Oil Company paved two public roads in Norway in 2007 with vegetable-oil-based asphalt. Results of this study are still premature.

HALIK Asphalts LTD from Israel has been experimenting with recycled and secondary road building since 2003. The company is using various wastes such as vegetable fats & oils, wax and thermoplastic elastomers to build and repair roads. The results reported are so far satisfying.

On October 6, 2010, a bicycle path in Des Moines, Iowa, was paved with bio-oil based asphalt through a partnership between Iowa State University, the City of Des Moines, and Avello Bioenergy Inc. Research is being conducted on the asphalt mixture, derived from plants and trees to replace petroleum-based mixes. Bioasphalt is a registered trademark of Avello Bioenergy Inc.

Dr. Elham H. Fini, at North Carolina A&T University, has been spearheading research that has successfully produced bio asphalt from swine manure.

References

- "National Grid: Half Britain's homes could be heated by renewable gas, says National Grid". Retrieved 11, June 2020
- "World's large river deltas continue to degrade from human activity". News Center. Retrieved 20, May 2020
- "Biogas Flows Through Germany's Grid Big Time - Renewable Energy News Article". 14 March 2012. Archived from the original on 14 March 2012. Retrieved 19, July 2020
- "ICIS Innovation Award Winners, Renmatix and Virent, Announce Collaboration On Bio-based Packaging", Retrieved 22, July 2020
- "GPS Renewables - Waste management through biogas - GPS Renewables". GPS Renewables. Retrieved 28, August 2020
- Sixta, Herbert, ed. (2006). Handbook of pulp. 1. Winheim, Germany: Wiley-VCH. pp. 79–88. ISBN 3-527-30997-7
- National Film Board of Canada. "Bate's Car: Sweet as a Nut". NFB.ca. Retrieved 03, June 2020
- "Cold climates no bar to biogas production". New Scientist. London: Sunita Harrington. 6 November 2010. p. 14. Retrieved 08, January 2020
- "Biogas plants provide cooking and fertiliser". Ashden Awards, sustainable and renewable energy in the UK and developing world. Retrieved 03, May 2020
- "404 - Seite nicht gefunden auf Server der Fachagentur Nachwachsende Rohstoffe e.V.: FNR" (PDF). Retrieved 24, March 2020
- Third Annual World Congress on Industrial Biotechnology and Bioprocessing, Toronto, ON, July 11–14, 2006 Retrieved 22, June 2020

- National Non-Food Crops Centre. "NNFCC Renewable Fuels and Energy Factsheet: Anaerobic Digestion", Retrieved 09, August 2020
- Elsevier Ltd, The Boulevard, Langford Lane, Kidlington, Oxford, OX5 1GB, United Kingdom. "Micro CHP report powers heated discussion about UK energy future". Retrieved 19, April 2020
- "Launch of New 'Ene-Farm' Home Fuel Cell Product More Affordable and Easier to Install - Headquarters News - Panasonic Newsroom Global". Retrieved 15, May 2020
- "Transforming Greenhouse Gas Emissions into Energy" (PDF). WIPO Green Case Studies, 2014. World Intellectual Property Organization. 2014. Retrieved 20, April 2020
- "McKinsey & Company Releases New Data About Growth in Industrial Biotech Sector at World Congress" Retrieved 03, July 2020

Permissions

All chapters in this book are published with permission under the Creative Commons Attribution Share Alike License or equivalent. Every chapter published in this book has been scrutinized by our experts. Their significance has been extensively debated. The topics covered herein carry significant information for a comprehensive understanding. They may even be implemented as practical applications or may be referred to as a beginning point for further studies.

We would like to thank the editorial team for lending their expertise to make the book truly unique. They have played a crucial role in the development of this book. Without their invaluable contributions this book wouldn't have been possible. They have made vital efforts to compile up to date information on the varied aspects of this subject to make this book a valuable addition to the collection of many professionals and students.

This book was conceptualized with the vision of imparting up-to-date and integrated information in this field. To ensure the same, a matchless editorial board was set up. Every individual on the board went through rigorous rounds of assessment to prove their worth. After which they invested a large part of their time researching and compiling the most relevant data for our readers.

The editorial board has been involved in producing this book since its inception. They have spent rigorous hours researching and exploring the diverse topics which have resulted in the successful publishing of this book. They have passed on their knowledge of decades through this book. To expedite this challenging task, the publisher supported the team at every step. A small team of assistant editors was also appointed to further simplify the editing procedure and attain best results for the readers.

Apart from the editorial board, the designing team has also invested a significant amount of their time in understanding the subject and creating the most relevant covers. They scrutinized every image to scout for the most suitable representation of the subject and create an appropriate cover for the book.

The publishing team has been an ardent support to the editorial, designing and production team. Their endless efforts to recruit the best for this project, has resulted in the accomplishment of this book. They are a veteran in the field of academics and their pool of knowledge is as vast as their experience in printing. Their expertise and guidance has proved useful at every step. Their uncompromising quality standards have made this book an exceptional effort. Their encouragement from time to time has been an inspiration for everyone.

The publisher and the editorial board hope that this book will prove to be a valuable piece of knowledge for students, practitioners and scholars across the globe.

Index

www.ingramcontent.com/pod-product-compliance
Lightning Source LLC
LaVergne TN
LVHW080805220926
848485LV00011B/209